**Practical Inductively Coupled
Plasma Spectrometry**

Practical Inductively Coupled Plasma Spectrometry

Second Edition

John R. Dean
Northumbria University
Newcastle-upon-Tyne
United Kingdom

This book is dedicated to my wife, Lynne and our two now grown-up children, Sam and Naomi, and our newest family member, Harris, the border terrier.

Each of you is pursuing your own goals and ambitions.

Contents

8.4.3   For ICP–AES   *189*
8.5   Useful Resources   *191*
           References   *198*
           Further Reading   *198*

9   Inductively Coupled Plasma: Troubleshooting and Maintenance   *201*
9.1   Introduction   *201*
9.2   Diagnostic Issues   *201*
9.3   Tips to Reduce...   *202*
9.3.1   Potential Autosampler Issues   *202*
9.3.2   Contamination   *202*
9.4   Tips to Improve...   *203*
9.4.1   Sample Preparation   *203*
9.5   How To ...   *203*
9.5.1   Unblock a Blocked Pneumatic Concentric Nebulizer   *203*
9.5.2   Clean the Spray Chamber   *203*
9.5.3   Clean the Plasma Torch   *204*
9.6   What to do About...   *204*
9.6.1   Plasma Ignition Problems   *204*
9.7   Shut Down Procedure (At the End of the Day)   *204*
9.8   Regular Maintenance Schedule   *205*

           The Periodic Table   *207*
           SI Units and Physical Constants   *209*
           Index   *213*

About the Author

John R. Dean DSc, PhD, DIC, MSc, BSc, FRSC, CChem, CSci, PFHEA

Since 2004, John R. Dean has been Professor of Analytical and Environmental Sciences at Northumbria University where he is also currently Head of Subject in Analytical Sciences, which covers all Chemistry and Forensic Science Programmes. His research is both diverse and informed, covering such topics as the development of novel methods to investigate the influence and risk of metals and persistent organic compounds in environmental and biological matrices, to development of new chromatographic methods for environmental and biological samples using gas chromatography and ion mobility spectrometry and development of novel approaches for pathogenic bacterial detection/identification. Much of the work is directly supported by industry and other external sponsors.

He has published extensively (over 200 papers, book chapters and books) in analytical and environmental science. He has also supervised over 30 PhD students.

John remains an active member of the Royal Society of Chemistry (RSC) and serves on several of its committee's including Analytical Division Council, Committee for Accreditation and Validation of Chemistry Degrees, Research Mobility Grant committee and is the International Coordinator for the Schools' Analyst Competition.

After a first degree in Chemistry at the University of Manchester Institute of Science and Technology (UMIST), this was followed by an MSc in Analytical Chemistry and Instrumentation at Loughborough University of Technology, and finally a PhD and DIC in Physical Chemistry at the Imperial College of Science and Technology (University of London). He then spent two years as a postdoctoral research fellow at the Government Food Laboratory in Norwich.

In 1988, he was appointed to a lectureship in Inorganic/Analytical Chemistry at Newcastle Polytechnic (now Northumbria University) where he has remained ever since.

In his spare time John is an active canoeist; he holds performance (UKCC level 3) coach awards in open canoe and white water kayak. In 2012, he was awarded an 'outstanding contribution' award by the British Canoe Union.

Preface

The technique of inductively coupled plasma (ICP) spectrometry has expanded and diversified in the form of a mini-revolution over the past 55 years. What was essentially an optical emission spectroscopic technique for trace element analysis has expanded into a source for both atomic emission spectrometry and mass spectrometry, capable of detecting elements at sub-ppb ($ng\,ml^{-1}$) levels with good accuracy and precision. Modern instruments have also shrunk in physical size, but expanded in terms of their analytical capabilities, reflecting the significant developments in both optical and semiconductor technology. Each of the nine chapters takes a particular aspect of the holistic field of ICP and outlines the key practical aspects.

In Chapter 1, information is outlined with regard to the general methodology for trace elemental analysis. This includes specific guidance on the potential contamination problems that can arise in trace elemental analysis, the basics of health and safety in the field and workplace and the practical aspects of recording a risk assessment. The focus of the chapter then moves to the numerical aspects of the topic with sections on units and appropriate assignment of the number of significant figures. Quantitative analysis requires an understanding and application of calibration graph plotting and interpretation. Numerical exercises involving the calculation of dilution factors and their use in determining original concentrations in aqueous and solid samples are provided as worked examples. Finally, the concept of quality assurance is introduced, together with the role of certified reference materials in trace element analysis.

Chapter 2 focuses on the specific area of sampling, sample storage and preservation techniques. Initially, however, the generic concepts of effective sampling are highlighted and contextualized. This is then followed by specific details on the sampling of soil, water and air. The major factors affecting sample storage are then addressed as well as practical remedies for the storage of sample. Finally, the possibilities for sample preservation are highlighted.

Chapter 3 considers the diverse of sample preparation strategies that have been adopted to introduce samples in to an ICP. These include the sample preparation approaches for the elemental analysis of metals/metalloids from solid and aqueous samples. The first part of this chapter is concerned with methods for the extraction of metal ions from aqueous samples. Emphasis is placed on liquid–liquid extraction, with reference to both ion-exchange and co-precipitation. The second part of this chapter is focused on the methods available for converting a solid sample into the appropriate form for elemental analysis. The most popular methods are based on acid digestion of the solid matrix, using either a microwave oven or a hot-plate approach. In addition, details are provided about the methods available for the selective extraction of metal species in soil studies using either single extraction, sequential extraction procedures or non-specific extraction. Finally, the role of in vitro simulated gastro intestinal and epithelial lung extraction procedures for estimated bioaccessibility are described.

Chapter 4 explores the different approaches available for the introduction of samples into an ICP. While the most common approach uses the generic nebulizer/spray chamber arrangement, the choice of which nebulizer and/or spray chamber requires an understanding of the principle of operation and the benefits of each design. Alternative approaches for discrete sample introduction are also discussed, including laser ablation while continuous sample introduction methods consider the coupling of flow injection and chromatography. Finally, opportunities for introducing gaseous forms of metals/metalloids using hydride generation and/or cold vapour techniques are discussed.

Chapter 5 describes the principle of operation of an ICP and the role of the radio frequency generator. The concept of viewing position is also introduced where it is of importance in atomic emission spectrometry where the plasma can be viewed either laterally or axially. In addition, the basic processes that occur within an ICP when a sample is introduced are discussed. Finally, a brief outline of the necessary signal processing and instrument control required for a modern instrument are presented.

Chapter 6 concentrates on the fundamental and practical aspects of inductively coupled plasma-atomic emission spectrometry (ICP–AES). After an initial discussion of the fundamentals of spectroscopy as related to atomic emission spectrometry, this chapter then focuses on the practical aspects of spectrometer design and detection. The ability to measure elemental information sequentially or simultaneously is discussed in terms of spectrometer design. Advances in detector technology, in terms of charge-transfer technology, are also highlighted in the context of ICP–AES.

Chapter 7 describes the fundamental and practical aspects of inductively coupled plasma–mass spectrometry (ICP–MS). After an initial discussion of the fundamentals of mass spectrometry, this chapter then focuses on the types of mass spectrometer and variety of detectors available for ICP–MS. The occurrence of isobaric and molecular interferences in ICP–MS is highlighted, along with suggested remedies. Of particular note is a discussion of collision and reaction cells in ICP–MS. Emphasis is also placed on the capability of ICP–MS to perform quantitative analysis using isotope dilution analysis (IDA).

Chapter 8 focuses on the current and future developments in ICP technology. After an initial comparison of ICP–AES and ICP–MS, the chapter considers the diversity of applications to which the technology has been applied. Finally, examples have been selected that highlight current and future developments for the ICP, in ICP–AES and ICP–MS. Some useful laboratory templates are also provided. This chapter concludes with guidance on the range of resources available to assist in the understanding of ICP technology and its application to trace element analysis.

Finally, Chapter 9 provides practical guidance on troubleshooting problems commonly encountered in the running of an ICP system. The chapter concludes by providing guidance on the maintenance schedule for maintaining an efficient and functioning ICP system.

John R. Dean
Northumbria University,
Newcastle, UK

Acknowledgements

This present text includes material that has previously appeared in several of the author's earlier books; that is, *Atomic Absorption and Plasma Spectroscopy* (ACOL Series, 1997), *Methods for Environmental Trace Analysis* (AnTS Series, 2003), *Practical Inductively Coupled Plasma Spectroscopy* (AnTS Series, 2005), *Bioavailability, Bioaccessibility and Mobility of Environmental Contaminants* (AnTS, 2007), *Extraction Techniques in Analytical Sciences* (AnTS, 2009) and *Environmental Trace Analysis: Techniques and Applications* (2014), all published by John Wiley & Sons, Ltd, Chichester, UK. The author is grateful to the copyright holders for granting permission to reproduce figures and tables from his two earlier publications.

In addition, the following are acknowledged.

Table 1.1 An example Control of Substances Hazardous to Health form. Reprinted with permission of Dr Graeme Turnbull, Northumbria University.

Figure 2.3 An illustration of a spring-loaded water sampling device. Reprinted with permission of Dynamic Aqua-Supply Ltd, Canada.
http://www.dynamicaqua.com/watersamplers.html

Figure 2.6 A schematic diagram of a high-volume sampler for collection of total suspended particulates. This work is licensed under the Creative Commons Attribution-ShareAlike 4.0 License. © CC BY 4.0 AU, Queensland Government, Australia.
https://www.qld.gov.au/environment/pollution/monitoring/air-pollution/samplers

Figure 3.11 Schematic layout of the human lungs system. Reproduced with permission of Humanbodyanatomy.co.
https://humanbodyanatomy.co/human-lungs-diagram/human-lungs-diagram-pictures-human-lungs-diagram-labeled-human-anatomy-diagram/

Figure 4.1 Schematic diagram for an autosampler sample presentation unit for ICP technology. Reproduced with permission of Elemental Scientific, Nebraska, USA.

http://www.icpms.com/products/brinefast-S4.php

Figure 4.2 Selected common commercially available nebulizers. This work is licensed under the Creative Commons Attribution-ShareAlike 4.0 License. © CC BY-SA 4.0, Burgener Research Inc.

Figure 4.10 A schematic diagram of a pneumatic concentric nebulizer – cyclonic spray chamber arrangement. Reproduced with permission of ThermoFisher.com.

https://www.thermofisher.com/de/en/home/industrial/spectroscopy-elemental-isotope-analysis/spectroscopy-elemental-isotope-analysis-learning-center/trace-elemental-analysis-tea-information/inductively-coupled-plasma-mass-spectrometry-icp-ms-information/icp-ms-sample-preparation.html

Figure 4.14 Schematic diagram of a laser ablation (LA) – ICP system. Reproduced with permission of Elsevier. Gunther, D., Hattendorf, B., Trends in Analytical Chemistry 24(3), (2015) 255–265.

Figure 4.17 A Schematic diagram of an HPLC - ICP system. Reproduced with permission of Elsevier. Delafiori, J., Ring, G., and Furey, A., Talanta 153, (2016) 306–331.

Figure 6.14 Schematic representation of the operation of a photomultiplier tube. Reproduced with permission of John Wiley and Sons Ltd., Chichester. Hou, X. and Jones, B.T., Inductively coupled plasma / optical emission spectrometry. Encyclopedia of Analytical Chemistry, Meyers, R.A. (Ed.), pp. 9468–9485.

Figure 6.16 Schematic representation of the operation of a charged-coupled device. Reproduced with permission of John Wiley and Sons Ltd., Chichester. Hou, X. and Jones, B.T., Inductively coupled plasma / optical emission spectrometry. Encyclopedia of Analytical Chemistry, Meyers, R.A. (Ed.), pp. 9468–9485.

Figure 7.4 Modeling the inductively coupled plasma temperature at the ICP-MS interface [5]. Reproduced with permission of the RSC. Bogaerts, A., Aghaei, M., J. Anal. At. Spectrom., 32, 233–261 (2017).

Figure 7.8 Schematic diagrams of the layout of high-resolution mass spectrometers for inductively coupled plasma (a) high resolution ICP-MS (HR-ICP-MS), sector field ICP-MS (SF-ICP-MS) or double-focusing ICP-MS (DF-ICP-MS) (e.g. a reverse geometry double-focusing magnetic sector MS), (b) ICP-tandem mass spectrometer (triple quadrupole ICP-MS or ICP-QQQ), and (c) multiple collector ICP-MS (MC-ICP-MS).

(a) Reproduced with permission of the RSC. Moldovan, M., Krupp, E.M., Holliday, A.E., Donard, O.F.X., J. Anal. At. Spectrom., 19, 815–822 (2004).

(b) Reproduced with permission of the RSC.

(c) Reproduced with permission of the Journal of Geostandards and Geoanalysis. Rehkamper, M., Schonbachler, M., and Stirling, C., Geostandards Newsletter, 25(1), 23–40 (2000).

Figure 7.10 Schematic representation of (a) ICP-tandem mass spectrometer (triple quadrupole ICP-MS or ICP-QQQ) and (b) its operating modes [11]. Reproduced with permission of the RSC. Bolea-Fernandez, E., Balcaen, L., Resano, M., and Vanhaecke, F., J. Anal. Atom. Spectrom., 32, 1660–1679 (2017).

Figure 7.11 Schematic representation of the operating principles for the different scanning options available for a ICP-tandem mass spectrometer (triple quadrupole ICP-MS or ICP-QQQ). (a) Product ion scan, (b) precursor ion scan, and (c) neutral mass gain scan [11]. Reproduced with permission of the RSC. Bolea-Fernandez, E., Balcaen, L., Resano, M., and Vanhaecke, F., J. Anal. Atom. Spectrom., 32, 1660–1679 (2017).

Figure 7.12 Schematic representation of the principle of overcoming spectral interferences for a ICP-tandem mass spectrometer (triple quadrupole ICP-MS or ICP-QQQ). (a) On-mass approach, (b) mass-shift approach (1), and (c) mass-shift approach (2) [11]. Reproduced with permission of the RSC. Bolea-Fernandez, E., Balcaen, L., Resano, M., and Vanhaecke, F., J. Anal. Atom. Spectrom., 32, 1660–1679 (2017).

Figure 7.13 An example of the principle of overcoming spectral interferences for a ICP-tandem mass spectrometer (triple quadrupole ICP-MS or ICP-QQQ) for the spectral free determination of ^{80}Se operated in MS/MS mode using both the (a) on-mass approach, and (b) mass-shift approach [11]. Reproduced with permission of the RSC. Bolea-Fernandez, E., Balcaen, L., Resano, M., and Vanhaecke, F., J. Anal. Atom. Spectrom., 32, 1660–1679 (2017).

Figure 7.17 Detectors for mass spectrometry: (a) a discrete dynode electron multiplier tube (EMT): mode of operation, (b) a continuous dynode (or channel) EMT: mode of operation, and (c) Faraday cup detector: mode of its operation. (c) Reproduced with permission from ThermoFisher Scientfic. https://www.thermofisher.com/uk/en/home/industrial/spectroscopy-elemental-isotope-analysis/spectroscopy-elemental-isotope-analysis-learning-center/trace-elemental-analysis-tea-information/inductively-coupled-plasma-mass-spectrometry-icp-ms-information/icp-ms-systems-technologies.html.

Figure 8.1 Photochemical vapour generation for ICP [1]. Reproduced with permission of RSC. Sturgeon, R.E., *J. Anal. At. Spectrom.*, 32, 2319–2340 (2017).

Figure 8.2 Schematic diagram of a (a) Flow Blurring® nebulizer and (b) its mode of operation. Reproduced with permission of Agilent com. https://www.agilent.com/en/products/mp-aes/mp-aes-supplies/mp-aes-oneneb-series-2-nebulizer.

Figure 8.3 Electrothermal vaporization (ETV) sample introduction device for an ICP [5]. Reproduced with permission of RSC. Hassler, J., Barth, P., Richter, S. and Matschat, R., *J. Anal. At. Spectrom.*, 26, 2404–2418 (2011).

Figure 8.6 Schematic diagram of a commercial ICP-ToF-MS [6]. Reproduced with permission of RSC. Hendriks, L., Gundlach-Graham, A., Hattendorf, B. and Gunther, D., *J. Anal. At. Spectrom.*, 32, 548–561 (2017).

Figure 8.7 Synchronous vertical dual-view of a commercial ICP-AES [7]. Reproduced with permission of RSC. Donati, G.L., Amais, R.S. and Williams, C.B., *J. Anal. At. Spectrom.*, 32, 1283–1296 (2017).

Acronyms, Abbreviations and Symbols

A_r	relative atomic mass
AC	alternating current
ACS	American Chemical Society
A/D	analogue-to-digital
AES	atomic emission spectrometry
ANOVA	analysis of variance
APDC	ammonium pyrrolidine dithiocarbamate
BEC	background equivalent concentration
C	coulomb
CCD	charge-coupled device; central composite design
CID	charge-injection device
COSHH	Control of Substances Hazardous to Health (Regulations)
CoV	coefficient of variation
CRM	Certified Reference Material
CTD	charge-transfer device
Da	dalton (atomic mass unit)
DC	direct current
DTPA	diethylenetriamine pentaacetic acid
EDTA	ethylenediamine tetraacetic acid
ESA	electrostatic analyser
ETV	electrothermal vaporization
eV	electron volt
GC	gas chromatography
HPLC	high performance liquid chromatography
Hz	hertz
IC	ion chromatography
ICP	inductively coupled plasma
ICP–AES	inductively coupled plasma–atomic emission spectrometry
ICP–MS	inductively coupled plasma–mass spectrometry
id	internal diameter
IDA	isotope dilution analysis

IUPAC	International Union of Pure and Applied Chemistry
J	joule
KE	kinetic energy
LC	liquid chromatography
LDR	linear dynamic range
LGC	Laboratory of the Government Chemist
LLE	liquid–liquid extraction
LOD	limit of detection
LOQ	limit of quantitation
M_r	relative molecular mass
MAE	microwave-accelerated extraction
MDL	minimum detectable level
MIBK	methylisobutyl ketone
MS	mass spectrometry
MSD	mass-selective detector
N	newton
NIST	National Institute of Standards and Technology
Pa	Pascal
PBET	physiologically based extraction test
PMT	photomultiplier tube
ppb	parts per billion (10^9)
ppm	parts per million (10^6)
ppt	parts per thousand (10^3)
RF	radiofrequency
RSC	Royal Society of Chemistry
RSD	relative standard deviation
SAX	strong anion exchange
SCX	strong cation exchange
SD	standard deviation
SE	standard error
SFMS	sector-field mass spectrometry
SI (units)	Système International (d'Unitès) (International System of Units)
TOF	time-of-flight
UV	ultraviolet
V	volt
W	watt
WWW	World Wide Web

c	speed of light; concentration
e	electronic charge
E	energy; electric field strength
f	(linear) frequency; focal length
F	Faraday constant

h	Planck constant
I	intensity; electric current
m	mass
M	spectral order
p	pressure
Q	electric charge (quantity of electricity)
R	resolution; correlation coefficient; molar gas constant; resistance
R^2	coefficient of determination
t	time; Student factor
T	thermodynamic temperature
V	electric potential
z	ionic charge
λ	wavelength
ν	frequency (of radiation)
σ	measure of standard deviation
σ^2	variance

1

The Analytical Approach

LEARNING OBJECTIVES

- To be aware of the different types of contamination that can cause problems in trace elemental analysis.
- To know about Health and Safety in the working environment.
- To be able to carry out a Risk Assessment for safe laboratory and in-field practice.
- To appreciate the units used in analytical chemistry.
- To be able to report numerical data with the appropriate assignment of significant figures.
- To be able to present numerical data with correct units and be able to interchange the units as required.
- To know how to present and report laboratory information in an appropriate format.
- To be able to determine the concentration of an element from a straight-line graph using the equation $y = mx + c$.
- To be able to calculate the dilution factor for a liquid sample and a solid sample, and hence determine the concentration of the element in the original sample.
- To appreciate the concept of quality assurance in the analytical laboratory.
- To be aware of the significance of certified reference materials in elemental analysis.
- To develop an understanding of reporting and interpreting the data generated in its appropriate context.

1.1 Introduction

Trace elemental analysis requires more than just knowledge of the analytical technique to be used; in this case, inductively coupled plasma spectrometry. It requires knowledge of a whole range of disciplines that need to come together to create the result. The disciplines required can be described as follows:

- Health and Safety in the laboratory (and external environment);

Practical Inductively Coupled Plasma Spectrometry, Second Edition. John R. Dean.
© 2019 John Wiley & Sons Ltd. Published 2019 by John Wiley & Sons Ltd.

- sampling, sample storage and preservation and sample preparation methodologies appropriate to sample type;
- analytical technique to be used;
- data control, including calibration strategies and the use of certified reference materials (CRMs) for quality control and data assurance;
- data management, including reporting of results and their interpretation, context and meaning.

While most of these are covered to some extent in this book, the reader should also consult other resources; for example, books, scientific journals and the web.

1.2   Essentials of Practical Work

The perspective that is required when faced with trace element analysis are the additional precautions required in terms of management of contamination, choice of reagents and acids and cleanliness of the workspace. For example, the grade of chemical used to prepare calibration standards is a major concern when working at (ultra) trace element analysis levels (sub-μg ml^{-1}). Chemicals are available in a range of grades from 'reagent grade' or 'technical grade' through to 'analytical reagent grade', for example, ACS reagent, AristaR®, >99% purity or PUROM™, Optigrade®, picograde® and ReagentPlus®. For 'analytical reagent grade' materials, the manufacturer has characterized the identity and concentration of impurities by subjecting it to stringent chemical analysis. Therefore, the use of sample and reagent blanks in the analytical procedure is essential to identify 'problem elements' that could interfere and to analyse and report data accordingly.

The risk of contamination is a major problem in trace element analysis. Apart from the analytical reagent used to prepare standards, as discussed previously, contamination can also be experienced from sample containers used for storage or quantitative analysis, for example, volumetric flasks, pipettes and so on. For example, metal ions can adsorb onto glass containers and then leach into the solution under acidic conditions, thereby causing contamination. This can be minimized by cleaning the glassware prior to use by soaking for at least 24 hours in a 10% nitric acid solution, followed by rinsing with 'clean' deionized water (three times). The cleaned vessels should then either be stored upside down or covered with Clingfilm® to prevent dust contamination.

For ultra-trace element analyses it will be necessary to perform all laboratory work in a cleanroom. Cleanrooms are designed to maintain extremely low levels of particulates, such as dust, airborne organisms or vaporized particles that could otherwise contaminate the sample or standard during its preparation and analysis. Ultimately, such contamination leads to an error in the reporting of the analytical result.

1.3   Health and Safety

It is a legal requirement for institutions to provide a working environment that is both safe and without risk to health. In the UK, the *Health and Safety at Work Act 1974* provides the legal framework for Health and Safety. The introduction of the *Control of Substances Hazardous to Health (COSHH)* Regulations in 2002 imposed specific legal requirements for risk assessment when hazardous chemicals (or biological agents) are used. This is often evidenced by the provision of training and information on safe working practices in the laboratory. For the student, this is often done by attending an appropriate safety briefing, the reading (and subsequent signing) of the safety booklet acknowledging an understanding of safety and their role in protecting themselves and other students as well as receiving appropriate training in the use of scientific equipment (e.g. the instrumentation described in this book).

Prior to undertaking any laboratory work, a *Risk Assessment* must be undertaken by an appropriately identified person (e.g. supervisor, academic or technical staff). The purpose of the Risk Assessment is to identify laboratory activities that could cause injury to people and then to provide control measures to ensure that the risk is reduced. Important considerations are:

- substance hazards
- how the substance is to be used
- how it can be controlled
- who is exposed
- how much exposure and its duration

It is important to distinguish between the hazard of a substance and its risk from exposure. This can be done by doing a *Risk Matrix Analysis (RMA)*. The RMA allows a prioritization of the *likelihood* and *severity* to the individual from the hazard identified. All manufacturers of hazardous chemicals are required to provide a *Material Safety Data Sheet (MSDS)* for the stated chemical. The MSDS will contain information about the chemical including:

- manufacturer
- name of chemical
- chemical components
- hazards associated with the product (including a Hazard Statement and a Precautionary Statement)
- first aid measures

- firefighting measures
- handling and storage
- accidental release procedures
- exposure control and personal protection
- physical and chemical properties
- stability and reactivity
- toxicological and ecological information
- disposal practices
- other miscellaneous information

With this information the user must then complete a COSHH form (Table 1.1). As part of the COSHH process specific details of the Hazard Statement and Precautionary Statement, for each chemical, must be included (Table 1.2). Then, assess the *likelihood* of harm coming to pass given the amount/nature of the chemical to be used and the environment/manner it is to be used in; at this stage, the likelihood is assessed on the basis that no specific control measures are being taken. The likelihood therefore assesses the highest risk. After assessing the likelihood, the next stage is to consider the *severity* of the risk. This is done by considering the substance-specific risk (rather than the activity specific risk). Again, like the likelihood this considers the highest severity. By then performing the RMA (*Risk = Likelihood × Severity*) (Table 1.3) you arrive at the risk for using the chemical.

The individual working in the laboratory is also a major source of contamination. Therefore, as well as the normal laboratory safety practices of wearing a laboratory coat and safety glasses, it may be necessary to take additional steps such as the wearing of 'contaminant-free' gloves and a close-fitting hat as well as working in a fume cupboard or for ultra-trace elemental analysis a cleanroom.

1.4 SI Units and Their Use

The Systeme International d'Unites (SI) uses a series of base units (Table 1.4) from which other terms have been derived. Some of the most commonly used SI derived units are shown in Table 1.5. When using units, it is standard practice to keep numbers between 0.1 and 1000 using a set of prefixes, based on multiples of 10^3 (Table 1.6). It is an extremely useful skill to be able to interchange these units and prefixes. For example, $1 \, mol \, l^{-1}$ can also be expressed as $1000 \, \mu mol \, ml^{-1}$, $1000 \, mmol \, l^{-1}$ or $1000 \, nmol \, \mu l^{-1}$. However, for practical purposes a $1 \, mol \, l^{-1}$ solution is the most useful term.

Table 1.1 An example Control of Substances Hazard to Health (COSHH) form.

Section 1: Overview

Names of chemicals to be used:	*Enter the name of each hazardous chemical to be used*
Title of activity:	*Enter the title of the activity*
Brief description of the activity:	*Briefly describe the activity to be undertaken*
Responsible person:	*Enter name of the member of staff responsible for your work e.g. supervisor*
Faculty / Department	*Enter the name of your Faculty / Department*
Date of assessment	*Enter the date* **Date of Re-assessment** *Enter the date one year from now*
Location of work:	*Enter the name of Building / Laboratory in which the work will be carried out.*

Section 2: Emergency Contacts (e.g. project supervisor).

Name	Position	Contact Telephone Number
Enter the name	*Enter their position*	*Enter their telephone number*

Section 3: Hazard Identification

3.1 For hazardous substances in this activity, click all that apply.					
	☐ Toxic		☐ Severe Health Hazards		☐ Health Hazards
	☐ Explosive		☐ Flammable		☐ Oxidising
	☐ Corrosive		☐ Gases Under Pressure		☐ Environmental

3.2 Select the hazard phrases (H-phrases) for each hazardous substance.			
1.	*Select a Hazard phrase*	e.g.	H302-Harmful if swallowed
e.g.	H226-Flammable liquid and vapour	e.g.	EUH014 - Reacts violently with water

(Continued)

Table 1.1 (Continued)

Section 4: Hazard Properties

Name of substance	Physical form	Quantity	Frequency	Route of exposure
Enter substance name	*Enter physical form*	*Enter the quantity*	*Enter the frequency*	*Select route*
e.g. Chemical name	solid dust	1 g	weekly	Ingestion

Section 5: Identifying Those at Risk

5.1 Who might be at risk? Select all that apply.		
☐ Staff/PGRs	☐ Taught Students	☐ Young persons (under 18 years old)
☐ New or expectant mothers	☐ Others:	

5.2 Assessment of risk to human health before control measures are in place.
Select the likelihood and severity of harm in the presence of the identified hazards **before** the control measures outlined above are implemented. Calculate the risk rating and act accordingly.

Likelihood of harm	Severity	Risk Rating and Outcome (likelihood x Severity)
e.g. 2. Unlikely	*2. Minor Injury*	*5. Good lab practice required.*

Section 6: Control Measures (Specify control procedures to each hazardous substance identified in section 4.)

6.1 Physical or engineering controls.				
☐ Laboratory	☐ Controlled area	☐ Total containment	☐ Glove Box	☐ Fume cupboard
☐ Microbial safety cabinet	☐ Local exhaust ventilation	☐ Access control	☐ Other: *Enter details*	
You must also specify below at which point in the work activity they are to be used.				
Specify at which point the control measures should be implemented				
6.2 Administrative controls.				
Describe administrative controls				

Table 1.1 (Continued)

	6.3 Personal protective equipment (PPE).		
	☐ Eyewear protection (Minimum standard CE EN166)		☐ Disposable lab coat
	☐ Lab coat		☐ Chemical suit
	☐ Specialised footwear. (Minimum standard EN ISO 20345) State type: *Enter details here*		☐ Hearing protection (Minimum standard EN352-1) State type: Enter details here
	☐ Gloves State minimum standard: ☐ BS EN455 – single use for chemical hazards. ☐ BS EN374 – single use for chemicals hazards and microorganisms. State type used: Enter details here		☐ Respirator State type: ☐ Disposable P3 (Minimum standard EN149) ☐ Replaceable filter (Minimum standard EN140) ☐ Powered respirator. State type used: *Enter details here*
	☐ Full-face visor		☐ Other State: *Enter details here*

6.4 Storage requirements.

Describe storage conditions

6.5 Transport of hazardous substances.

Describe how you will transport the hazardous substances

6.6 Disposal of waste. If specialised waste is to be generated, you must discuss this with a member of technical staff and consult the university waste policy.

Waste type	Waste subtype	Detail method of disposal
Select waste type e.g. liquid	*Select a waste sub-type e.g. Inorganic waste*	*Describe the method of disposal e.g. down the sink with plenty of water*

6.7 Emergency procedures.

Minor spillage (for less than 250 mL / 250 g of materials with a low-medium risk rating).	Major spillage (for greater than 250 mL / 250 g of materials with a low-medium risk rating, or <u>any</u> high risk materials).
☐ Secure the spill area.	☐ Evacuate and secure the laboratory/area.
☐ Inform a competent person (e.g. a member of technical staff or your supervisor).	☐ Inform a competent person (e.g. a member of technical staff or your supervisor).

(Continued)

Table 1.1 (Continued)

☐ Other *Describe other emergency procedures*	☐ Evacuate the building using the fire alarm.

In the event of fire, assuming you are trained in the handling of extinguishers <u>and</u> it is safe to do so, specify which types of fire control may be used:

☐ Carbon dioxide	☐ Water	☐ Dry powder	☐ Foam	☐ Fire blanket	☐ Automatic fire suppression
☐ Other	*Describe other fire control measures here, if applicable*				

In the event of an accident requiring first aid, seek assistance as soon as possible.
Detail below any specific considerations, which must be made for the hazardous substances in use.

☐ If hazardous material comes into contact with skin, remove any affected clothing and wash the area with copious amounts of water. ☐ For large areas rinse the skin using the emergency shower.	☐ If hazardous material comes into contact with the eyes, rinse the eyes using an eye wash station. ☐ For serious eye burns, use diphoterine station.
☐ For phenol burns, wash with copious amounts of water and apply polyethylene (PEG) 300 to the area.	☐ For hydrofluoric acid burns, wash with copious amounts of water and apply calcium gluconate gel to the area.
☐ If cyanide has been inhaled, move the victim to fresh air.	☐ Other. Please state: *Enter details here*

6.8 <u>Assessment of risk</u> to human health once control measures are in place.
Select the likelihood and severity of harm in the presence of the identified hazards after the control measures outlined above are implemented. Calculate the risk rating and act accordingly. Guidance may be found in the appendix by clicking <u>here</u>.

Likelihood of harm	Severity	Risk Rating and Outcome (Likelihood x Severity)
Enter likelihood e.g. 2. Unlikely	*Enter severity e.g. 1. Delay only*	*Enter risk rating e.g. 2. Good laboratory practice only required*

6.9 Instruction, training and supervision.
In consultation with the approver, specify the level of training and supervision required to safely carry out the work described. **Select all that apply.**

☐ Special instructions are required to safely carry out the work.	
☐ Special training is required to safely carry out the work.	
☐ Work may be carried without direct supervision.	☐ Work may be carried without indirect supervision.
☐ Work may not be started without the advice and approval of the approver.	☐ Work may not be carried out without close supervision.

Table 1.1 (Continued)

Section 7: Approval
I hereby confirm that the above is a suitable and sufficient risk assessment for the work activity described.

7.1 The assessor.		
Name	Signature	Date
7.2 The approver (if required).		
Name	Signature	Date

Table 1.2 Examples of (a) Hazard[a] and (b) Precautionary[b] statements.

(a)

Letter	Type of hazard	Intrinsic properties of the substance	Example
H	2 = Physical	e.g. Explosive properties for codes 200–210; flammability for codes 220–230 etc.	H302 harmful if swallowed
H	3 = Health		
H	4 = Environmental		

(b)

Letter	Type of precaution	Examples
P	1 = General precaution	P102 Keep out of the reach of children
P	2 = Prevention precaution	P281 Use personal protective equipment as required
P	3 = Response precaution	P301 If swallowed:
P	4 = Storage precaution	P404 Store in a closed container
P	5 = Disposal precaution	P501 Dispose of contents/container to ...

a) There are 72 individual and 17 combined Hazard statements.
b) There are 116 individual and 33 combined Precautionary statements.

Table 1.3 Risk matrix analysis.[a]

		Severity					
		6	5	4	3	2	1
		multiple fatalities	single fatality	major injury	lost time injury	minor injury	delay only
6	certain	36	30	24	18	12	6
5	very likely	30	25	20	15	10	5
4	likely	24	20	16	12	8	4
3	may occur	18	15	12	9	6	3
2	unlikely	12	10	8	6	4	2
1	remote	6	5	4	3	2	1

Likelihood (row label)

a) *Note: Low risk:* numerical score 1–10. Good laboratory practice (including Personal Protective Equipment of a laboratory coat and safety glasses) required. *High risk:* numerical score 12–18. Specific identified control measures must be used. *Very high risk:* numerical score 20+. Trained personnel only.

Table 1.4 Some commonly used base SI units.

Measured quantity	Name of SI unit	Symbol
Length	Metre	m
Mass	Kilogram	kg
Amount of substance	Mole	mol
Time	Second	s
Thermodynamic temperature	Kelvin	K

Table 1.5 Some commonly used derived SI units.

Measured quantity	Name of unit	Symbol	Definition in base units	Alternative in derived units
Electric charge	Coulomb	C	$A\,s$	$J\,V^{-1}$
Energy	Joule	J	$m^2\,kg\,s^{-2}$	$N\,m$
Force	Newton	N	$m\,kg\,s^{-2}$	$J\,m^{-1}$
Frequency	Hertz	Hz	s^{-1}	—
Pressure	Pascal	Pa	$kg\,m^{-1}\,s^{-2}$	$N\,m^{-2}$
Power	Watt	W	$m^2\,kg\,s^{-3}$	$J\,s^{-1}$

Table 1.6 Commonly used prefixes.

Multiple	Prefix	Symbol
10^{15}	peta	P
10^{12}	tera	T
10^9	giga	G
10^6	mega	M
10^3	kilo	k
10^{-3}	milli	m
10^{-6}	micro	μ
10^{-9}	nano	n
10^{-12}	pico	p
10^{-15}	femto	f

1.5 Significant Figures

A common issue when recording data from practical work is the mis-reporting of significant figures. This issue is important as it conveys an understanding of the key concepts in data treatment. The following examples illustrate the issues and how they can be interpreted.

For example, when asked to accurately weigh an approximate 0.5 g of sample, how many decimal places should be reported? In this situation it would be expected that a four-figure decimal place analytical balance would be used to accurately weigh out the sample. On that basis, the sample would be recorded as 0.5127 g.

In practical terms, the sample would have been weighed by difference, that is, a sample container would be first weighed on a four-figure decimal place analytical balance, then the sample placed inside the container and the weight again recorded and, finally, the sample transferred to a digestion vessel and the sample container re-weighed. By taking the weights of the container with/without the sample allows an accurate recording of the weight of sample transferred into the digestion vessel.

In addition, for example, if you have a numerical value, representing a weight or concentration, of 276.643 it would be reasonable to represent this as: 276.6 or even 277. If the value was 0.828, then it may be reasonable to round up to 0.83. Whereas for a value of 12 763. It would be reasonable to report as 12 763 or, in some circumstances, 12 760.

Finally, for example, is it appropriate to report the concentration of an element in a solid sample as 25.21345678 mg kg^{-1}? No, a more appropriate reporting of the concentration would be 25.2 mg kg^{-1}.

In general terms the following guidance is provided:

- When rounding up numbers to the fourth decimal place, add one to the last figure if the number is greater than 5; for example, 0.54667 would become 0.5467.
- When rounding down numbers to the fourth decimal place, remove one to the last figure if the number is less than 5; for example, 0.54662 would become 0.5466.
- For a number 5, round to the nearest even number; for example, 0.955 would become 0.96 (to two significant figures) *or* if the value before 5 is even, it is left unchanged; for example, 0.945 would become 0.94 (to two significant figures) *or* if the value before 5 is odd, its value is increased by one; for example, 0.955 would become 0.96 (to two significant figures).
- Zero is not a significant figure when it is the first figure in a number; for example, 0.0067 (this has two significant figures, 6 and 7). In this situation it is best to use scientific notation, for example, 6.7×10^{-3} or, as the number is normally associated with a unit use the prefix milli, m; for example, 6.7 mg.

1.6 Calibration and Quantitative Analysis

Quantitative analysis in plasma spectroscopy requires the preparation of a series of calibration standards from a stock solution. These standards are prepared in volumetric flasks. Calibration solutions are usually prepared in terms of their molar concentrations, that is, $mol\,l^{-1}$, or mass concentrations, that is, $g\,l^{-1}$, with both referring to an amount per unit volume; that is, concentration = amount/volume. It is important to use the highest (purity) grade of chemicals (liquids or solids) for the preparation of the stock solution; for example, an analytical reagent grade.

For example, the preparation of a $0.1\,mol\,l^{-1}$ solution of lead from its metal salt in a 1 l volumetric flask would be done as follows. Using the molecular weight of lead nitrate, $Pb(NO_3)_2$ (331.20) and the atomic weight of lead (207.19) you simply multiply the molecular weight by the desired molarity ($0.1\,mol\,l^{-1}$) to give you the exact amount of lead nitrate to be dissolved in 1 l of solution to produce a $0.1\,mol\,l^{-1}$ solution of lead.

$$\text{that is, } (331.20\ \text{g mol}^{-1} \times 0.1\ \text{mol l}^{-1}) = 33.1200\ \text{g of } Pb(NO_3)_2 \text{ in } 1\ l$$

However, it is often the case that 1 l of solution would not be required; a more realistic volume would be 100 ml. In that case, each weight of material would need to be divided by 10. Therefore, 3.3120 g of $Pb(NO_3)_2$ would be dissolved in 100 ml of solution to produce a $0.1\,mol\,l^{-1}$ Pb solution.

Similarly, for example, if you wish to prepare a $1000\,\mu g\,ml^{-1}$ solution of lead from lead nitrate you simply divide the molecular weight by the atomic weight to give you the exact amount of lead nitrate to be dissolved in 1 l of solution to produce a $1000\,\mu g\,ml^{-1}$ solution of lead.

that is, $(331.20\,/\,207.19) = 1.5985\,g$ of $Pb(NO_3)_2$ in 1 l.

However as before, it is often the case that 1 l of solution would not be required; a more realistic volume would be 100 ml. In that case each weight of material would need to be divided by 10. Therefore, $0.1599\,g$ of $Pb(NO_3)_2$ would be dissolved in 100 ml of solution to produce a $1000\,\mu g\,ml^{-1}$ stock solution of Pb.

[*Note:* the mass concentration $\mu g\,ml^{-1}$ (it could also be expressed as $mg\,l^{-1}$, for example) is also referred to as ppm (parts per million).]

1.7 Making Notes of Practical Work and Observations

It is often convenient when carrying out laboratory work to record your data in a laboratory notebook (hardback paper or electronic notebook format). It is important that you always accurately record, in whatever format used, your name, module code, details of what you did, reagents and chemicals used, equipment and instrumentation used, accurate weights and volumes, duration times and practical observations. Some specific tips are provided on recording information in your hardback notebook:

- Record data correctly and legibly (even you may not be able to read your own writing later).
- Write in ink (and not pencil, which fades with age).
- Include the date and title of individual experiments and/or areas of investigation.
- Briefly outline the purpose of the experiment; that is, what you hope to know by the end.
- Identify and record the hazards and risks associated with the chemicals/equipment being used.
- Refer to the method/procedure being used (undergraduate laboratory) or write a full description of the method/procedure and its origins (postgraduate research).
- Record your observations (and note your interpretation at this stage); for example, accurate weights, volumes, how standards and calibration solutions were prepared and instrumentation settings (and the actual operating parameters).

Table 1.7 Recording quantitative data for the analysis of lead by inductively coupled plasma–mass spectrometry (ICP–MS).

Concentration (μg l^{-1})	^{208}Pb intensity (counts s^{-1})
0	565
10	19 887
20	45 356
30	59 876
40	78 543
50	99 654

- Record data with the correct units, for example, mg or μg g^{-1}, and to an appropriate number of significant figures, for example, 26.3 mg or 0.48 μg g^{-1} (and not 26.3423 mg or 0.4837 μg g^{-1}).
- Interpret data in the form of tables and calibration graphs.
- Record initial conclusions.

[*Note:* Example templates for the recording of laboratory information are provided in Chapter 8. Included in the Chapter 8 Appendices are example templates for: Sample pre-treatment; Sample preparation; ICP–AES analysis and ICP–MS analysis.]

A useful approach is to accurately record quantitative laboratory results in tabular format. This is often best done by creating two columns into which the data can be entered. It is essential, however, for future consultation of these data, that the columns are given the appropriate headings, for example, concentration (μg l^{-1}) and signal (mV), to prevent errors occurring later. It is also important to record details of any sample dilutions that have taken place (see Chapter 8, Laboratory Templates). A typical table of data for an experiment to determine the concentration of lead is shown in Table 1.7. In the case of electronic (e)-notebooks laboratory proformas will be available, as per the Laboratory Templates in Chapter 8, that allow information and data to be inserted and recorded in a pre-determined format and style.

1.8 Data Analysis

In quantitative data analysis it is normal to plot a graph using either dedicated instrument software or another commercial computer-based graphics package, for example, Microsoft Excel™, rather than by hand on graph paper. [*Note:* You should remember, however, that you may still need the skill to plot a graph by hand on graph paper in a university examination unless the examination has gone on-line!] Irrespective of the mode of preparing the graph, it is important to ensure that the graph is correctly labelled and presented. All graphs should have a numerical descriptor and title, for example, 'Figure 5.1 Calibration of Pb by ICP-AES'.

Graphs are normally used to describe a relationship between two variables, for example, x and y. It is normal practice to identify the x-axis as the horizontal axis (abscissa) and to use this for the independent variable, for example, concentration (with its appropriate units, e.g. $\mu g\,ml^{-1}$). The y-axis as the vertical axis (ordinate) is used to plot the dependent variable, for example, signal response (with appropriate units, e.g. mV). The mathematical relationship used for linear straight-line graphs is:

$$y = mx + c \tag{1.1}$$

where y is the signal response, for example, signal (mV), x is the concentration of the working solution (in appropriate units, e.g. $\mu g\,ml^{-1}$), m is the slope of the line of best fit of the graph and c is the intercept on the x-axis.

Then, by a simple re-arrangement allows the determination of the unknown sample concentration (x):

$$(y - c)/m = x \tag{1.2}$$

A typical graphical representation of data (a *direct calibration graph*) obtained from an experiment to determine the level of lead in a sample using inductively coupled plasma–mass spectrometry (ICP–MS) is shown in Figure 1.1a.[1]

An alternative approach to undergoing a *direct calibration*, as described before, is the use of the method of *standard additions*. This may be particularly useful if the sample is known to contain a significant portion of a potentially interfering matrix. In standard additions, the calibration plot no longer passes through zero (on both the x- and y-axes). As the concept of standard additions is to eliminate any matrix effects present in the sample, it should be implicit that the working standard solutions will all contain the same volume of the sample; that is, the same volume of the sample solution is introduced into a succession of working calibration solutions. Each of these solutions containing the same volume of the sample is then introduced into the inductively coupled plasma and the response recorded. However, plotting the signal response (e.g. signal (mV)) against analyte concentration produces a graph that no longer

---

1 R is known as the *correlation coefficient*, and provides a measure of the quality of calibration. In practice, R^2 (the *coefficient of determination*) is used because it is more sensitive to changes. This varies between -1 and $+1$, with values very close to -1 and $+1$ pointing to a very tight 'fit' of the calibration curve.

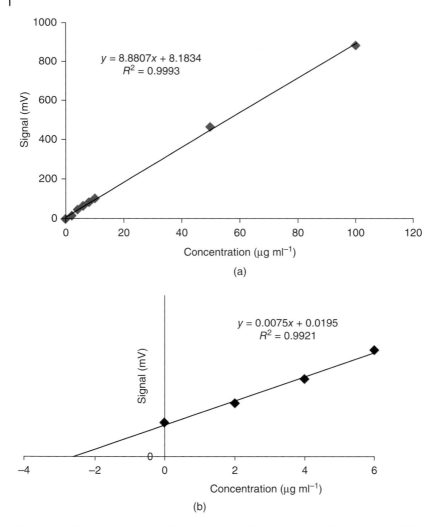

Figure 1.1 Calibration graphs. (a) a direct calibration graph and (b) a standard additions method calibration graph.

passes through zero on either axis, but if correctly drawn, the graph can be extended towards the x-axis (extrapolated) until it intercepts it. By maintaining a constant concentration x-axis, the unknown sample concentration can be determined (Figure 1.1b).[1] It is essential that this graph is linear over its entire length or otherwise considerable error can be introduced.

The limit of detection (LOD) of an analytical procedure is the lowest amount of analyte in an unknown sample that can be detected but not necessarily quantified, that is, recorded as an exact concentration. The LOD, expressed as a concentration (in appropriate units), is derived from the smallest measure, x, that can be detected with reasonable certainty for a given procedure. One approach to determine the LOD is to measure the signal of a known concentration at or near the lowest concentration that is observable (normally at least seven times). The value C_L is given by the equation:

$$C_L = C_{LCS} + K.SD_{LCS} \tag{1.3}$$

where C_{LCS} is the mean of the low concentration standard, SD_{LCS} is the standard deviation of the low concentration standard and K is a numerical factor chosen according to the confidence level required (typically 2 or 3).

An alternate format, that uses the signal-to-background ratio (SBR) to calculate the LOD is:

$$C_L = (3 \times C_{LCS} \times RSD_B) / SBR \tag{1.4}$$

where RSD_B is the relative standard deviation of the background signal.

As LODs are often not practically measurable a more realistic value is to use the limit of quantitation (LOQ) of an analytical procedure. The LOQ is the lowest amount of a metal in a sample that can be quantitatively determined with suitable certainty; the LOQ can be taken as 10 times 'the signal-to-noise ratio' or $K = 10$ in Eq. (1.3).

1.9   Data Treatment

In both the direct calibration graph and the method of additions graph, the result obtained from the sample is normally not the final answer of how much of the metal was in the original sample. This is because the sample has normally undergone some form of sample preparation (see Chapter 3). In the case of a solid sample, this might have involved acid digestion (see Section 3.2.1), while in the case of a liquid sample, liquid–liquid extraction (see Section 3.1.1) or other form of extraction (see Section 3.3). What is therefore required is a correction to the concentration data obtained from the calibration; often the application of a dilution or concentration factor that considers the sample preparation procedure. The following provides examples of the general forms of calculations that are necessary in the case of (i) a liquid sample that has been extracted using ammonium pyrrolidine dithiocarbamate (APDC)–methyl isobutyl ketone (MIBK) and (ii) a solid sample that has been acid-digested or extracted.

For example, calculate the concentration ($\mu g\,ml^{-1}$) of copper in a waste water sample obtained from the local waste treatment plant. A waste water sample (150 ml) was extracted with APDC–diethylammonium diethyldithiocarbamate (DDDC) into MIBK (20 ml) using liquid–liquid extraction. The extract was then quantitatively transferred to a 25.0 ml volumetric flask and made up to the mark with MIBK. What is the dilution factor?

$$25 \, ml / 150 \, ml = 0.167 \, ml \, ml^{-1} = 0.167 \text{ (with no units)}$$

[*Note:* that the dilution factor only considers the final volume of extract (i.e. 25.0 ml) and the initial sample volume (150 ml of waste water).]

If the solution was then analysed and found to be within the linear portion of the graph (see Figure 1.1a), the value for the dilution factor should then be multiplied by the concentration from the graph, producing a final value indicating the concentration of copper in the waste water sample.

In addition, for example, calculate the concentration ($\mu g \, g^{-1}$) of lead in a soil sample obtained from a contaminated land site. An accurately weighed (5.2456 g) soil sample was acid-digested using nitric acid and hydrogen peroxide, cooled and then quantitatively transferred to a 100.0 ml volumetric flask and made up to the mark with distilled water. This solution was then diluted by taking 10.0 ml and transferring to a further 100.0 ml volumetric flask where it is made up to the mark with high-purity water. What is the dilution factor?

$$[(100.0 \, ml) / (5.2456 \, g) \times (100.0 \, ml / 10.0 \, ml)] = 190.64 \text{ ml } g^{-1}$$

If the solution was then analysed and found to be within the linear portion of the graph (see Figure 1.1a), the value for the dilution factor should then be multiplied by the concentration from the graph, so producing a final value indicating the concentration of lead in the contaminated soil sample.

[*Note*: This type of calculation would be used for the use of alternate extraction protocols (see Section 3.3).]

1.10 Data Quality

It is essential to know whether the data obtained is appropriate. Quality assurance is all about getting the correct result. The main objectives of a quality assurance scheme are as follows:

- to select and validate appropriate methods of sample preparation;
- to select and validate appropriate methods of analysis;
- to maintain and upgrade analytical instruments;
- to ensure good record-keeping of methods and results;
- to ensure quality of the data produced;
- to maintain a high quality of laboratory performance.

The following are examples of important aspects of establishing and maintaining such a QA scheme:

- Individual performing the analyses
  - Has the individual been trained in the use of the instrumentation and/or procedures? If so by whom (where they trained or experienced)?
  - Was the training formal (formal qualification or certificate of competency obtained) or done in-house?
  - Can the individual use the instrumentation alone or do they require oversight?

- Laboratory procedures and practices
  - Do the procedures use CRMs to assess the accuracy of the method?
  - Do the procedures use spiked samples to assess recoveries? Samples are spiked with a known concentration of the analyte under investigation and their recoveries noted; this allows an estimate of analyte matrix effects to be made.
  - Do the procedures include analysis of reagent blanks? Always analyse reagents whenever the batch is changed or a new reagent introduced. This allows reagent purity to be assessed and, if necessary controlled, and also acts to assess the overall procedural blank; typically introduce a minimum number of reagent blanks; that is, 5% of the sample load.
  - Do the procedures use standards to calibrate instruments? A minimum number of standards should be used to generate the analytical curve, for example, a minimum of five. Daily verification of the calibration plot should be done using one or more standards within the linear working range.
  - Do the procedures include the analysis of duplicate samples? Analysis of duplicates or triplicates allows the precision of the method to be determined and reported.
  - Do the procedures include known standards within the sample run? A known standard should be run after every 10 samples to assess instrument stability; this also verifies the use of a daily calibration plot.

In implementing a good quality control programme, it is necessary to analyse a CRM. A CRM is a substance for which one or more elements have known values and estimates of their uncertainties, produced by a technically valid procedure, accompanied with a traceable certificate and issued by a certifying body. Typical examples of certifying bodies are the National Institute for Standards and Technology (NIST), based in Washington, D.C., USA, the Community Bureau of Reference (BCR), Brussels, Belgium and the Laboratory of the Government Chemist (LGC), London, UK. The accompanying certificate, in addition to providing details of the certified elemental concentration and their uncertainties in the sample, also provides details of the minimum sample weights to be used, storage conditions, moisture content and so on. An example of a typical certificate is shown in Figure 1.2. The incorporation of a suitable CRM alongside your unknown samples provides an opportunity for the accuracy (see Section 1.12) of your sample preparation methodology and analysis protocol to be investigated. A large range of CRMs across a broad range of sample matrices are available. It is appropriate to choose a CRM with a similar/same matrix as your sample types. Then, by comparing your obtained element concentration data for the CRM and its certificate value you can decide on whether your sample preparation methodology and analysis protocol are appropriate (or not). Agreement with the element concentration data to within a standard deviation of the certificate data confirms its suitability for application with the unknown samples. If the values obtained are outside the defined certified data range, a re-evaluation of *all* procedures, working practices and reagents used is required to establish any inherent issue(s) that have led to inaccurate data being obtained. Once the situation is resolved, and often this is by trial and error as well as intuition and past experience, the CRM would be re-sampled, prepared and re-analysed until the data obtained is within the defined data limits for each element.

Figure 1.2 Diagrammatic representation of a certificate as supplied with a Certified Reference Material.

International Organization Name			
Type of matrix			
Element	Concentration (wt.%)	Element	Concentration (mg kg^{-1})
Calcium	1.504 ± 0.013	Cadmium	(0.011)*
Magnesium	0.251 ± 0.009	Copper	5.34 ± 0.21
Phosphorus	0.140 ± 0.009	Lead	0.560 ± 0.022
Potassium	1.31 ± 0.03	Nickel	0.86 ± 0.09
Sulfur	(0.09)*	Zinc	12.1 ± 0.4

Notes:

The material will be provided in a sealed container alongside this certificate. The certificate, in addition to providing details of the certified (and indicative) elemental concentrations and their uncertainties in the sample, also provides details of the minimum sample weights to be used, storage conditions, moisture content, details of how analysed and so on.

*Values in parentheses are indicative values only.

1.11   Data Interpretation and Context

It is essential that the data obtained is placed in its context and interpreted. For example, the concentration obtained by the analysis and verified by a quality assurance scheme generates a trustworthy and reliable value. However, the interpretation and contextualization of the obtained metal concentration is essential. This might be, for example, against known legislative values that need to be enforced, for food protection and dietary requirements, metallurgical analysis of steel for bridge building and metal impurities in a catalyst for polymer production. So, it might be that the concentration obtained in the original sample is reported or that the concentration obtained is converted into a decision, for example, safe or not safe, good quality or poor quality.

1.12   Analytical Terms and Their Definitions

Finally, some useful analytical terms and their definitions are presented. The most important analytical terms of use in practical inductively coupled plasma spectrometry are:

- *Accuracy.* A quantity referring to the difference between the mean of a set of results or an individual result and the value that is accepted as the true or correct value for the quantity being measured.
- *Acid digestion.* Use of acid (and often heat) to destroy the organic matrix of a sample to liberate the metal content.
- *Aliquot.* A known amount of a homogenous material assumed to be taken with negligible sampling error.
- *Analyte.* The component of a sample that is ultimately determined directly or indirectly.
- *Bias.* Characterizes the systematic error in each analytical procedure and is the (positive or negative) deviation of the mean analytical result from the (known or assumed) true value.
- *Calibration.* The set of operations that establish, under specified conditions, the relationship between values indicated by a measuring instrument or measuring system and the corresponding known values of the measurand.
- *Calibration curve.* Graphical representation of a measuring signal as a function of quantity of analyte.
- *Certified Reference Material (CRM).* Reference material, accompanied by a certificate, one or more of whose property values are certified by a procedure that establishes its traceability to an accurate realization of the units in which the property values are expressed, and for which each certified value is accompanied by an uncertainty at a stated level of confidence.

- *Coefficient of determination.* The measure of the quality of calibration is often expressed as R^2 because it is more sensitive to changes. Values vary between −1 and +1, with values very close to −1 and +1 pointing to a very tight 'fit' of the calibration curve.
- *Complexing agent.* The chemical species (an ion or a compound) that will bond to a metal ion using lone pairs of electrons.
- *Confidence interval.* Range of values that contains the true value at a given level of probability. The latter is known as the *confidence level.*
- *Confidence limit.* The extreme values or 'end-values' in a confidence interval.
- *Contamination.* In trace analysis this is the unintentional introduction of analyte(s) or other species that are not present in the original sample and may cause an error in the determination. This can occur at any stage in the analysis. Quality assurance procedures, such as analyses of blanks or of reference materials, are used to check for contamination problems.
- *Control of Substances Hazardous to Health (COSHH).* Regulations that impose specific legal requirements for risk assessment wherever hazardous chemicals (or biological agents) are used.
- *Co-precipitation.* The inclusion of otherwise soluble ions during the precipitation of lower-solubility species.
- *Correlation coefficient.* The measure of the quality of calibration (R). R^2 is known as the *coefficient of determination.*
- *Dilution factor.* The mathematical factor applied to the determined value (data obtained from a calibration graph) that allows the concentration in the original sample to be determined. Frequently, for solid samples, this will involve a sample weight and a volume to which the digested/extracted sample is made up to prior to analysis. For liquid samples, this will involve an initial sample volume and a volume to which the digested/extracted sample is made up to prior to analysis.
- *Dissolved.* Material that will pass through a 0.45 µm membrane filter assembly prior to sample acidification.
- *Dry ashing.* Use of heat to destroy the organic matrix of a sample to liberate the metal content.
- *Error.* The error of an analytical result is the difference between the result and a 'true' value:
  - *Random error.* Result of a measurement minus the mean that would result from an infinite number of measurements of the same measurand carried out under repeatability conditions.
  - *Systematic error.* The mean that would result from an infinite number of measurements of the same measurand carried out under repeatability conditions, minus the true value of the measurand.

- *Extraction.* The removal of a soluble material from a solid mixture by means of a solvent or the removal of one or more components from a liquid mixture by use of a solvent with which the liquid is immiscible or nearly so.
- *Figure of merit.* A parameter that describes the quality of performance of an instrument or an analytical procedure.
- *'Fitness for purpose'.* The degree to which data produced by a measurement process enables a user to make technically and administratively correct decisions for a stated purpose.
- *Heterogeneity.* The degree to which a property or a constituent is randomly distributed throughout a quantity of material. The degree of heterogeneity is the determining factor of sampling error.
- *Homogeneity.* The degree to which a property or a constituent is uniformly distributed throughout a quantity of material. A material may be homogenous with respect to one analyte but heterogeneous with respect to another.
- *Interferent.* Any component of the sample affecting the final measurement.
- *Limit of detection (LOD).* The detection limit of an individual analytical procedure is the lowest amount of an analyte in a sample that can be detected but not necessarily quantified as an exact value. The LOD, expressed as either the concentration C_L or the quantity Q_L, is derived from the smallest measure, C_L, that can be detected with reasonable certainty for a given procedure. The value C_L is given by Eq. (1.3). For many purposes, the LOD is taken to be $3S_{bl}$ or 3 times 'the signal-to-noise ratio', assuming a zero blank.
- *Limit of quantitation.* For an individual analytical procedure, this is the lowest amount of an analyte in a sample which can be quantitatively determined with suitable uncertainty. It may also be referred to as the *limit of determination*. The LOQ can be taken as 10 times 'the signal-to-noise ratio', assuming a zero blank.
- *Linear dynamic range (LDR).* The concentration range over which the analytical working calibration curve remains linear.
- *Linearity.* This defines the ability of the method to obtain test results proportional to the concentration of analyte.
- *Liquid–liquid extraction.* A method of extracting a desired component from a liquid mixture by bringing the solution into contact with a second liquid, the solvent, in which the component is also soluble, and is immiscible with the first liquid or nearly so.
- *Matrix.* The carrier of the test component (analyte), all of the constituents of the material except the analyte, or the material with as low a concentration of the analyte as it is possible to obtain.
- *Measurand.* A quantity subject to measurement.
- *Method.* The overall, systematic procedure required to undertake an analysis. This includes all stages of the analysis, and not just the (instrumental) end determination.

- *Microwave digestion.* A method of digesting an organic matrix to liberate metal content by using an acid at elevated temperature (and pressure) based on microwave radiation. Can be carried out in either open or sealed vessels.
- *Organometallic.* An organic compound in which a metal is covalently bonded to carbon.
- *Outlier.* This may be defined as an observation in a set of data that appears to be inconsistent with the remainder of that set.
- *Precision.* The closeness of agreement between independent test results obtained under stipulated conditions.
- *Qualitative analysis.* Chemical analysis designed to identify the components of a substance or mixture.
- *Quality assurance.* All those planned and systematic actions necessary to provide adequate confidence that a product or services will satisfy given requirements for quality.
- *Quality control.* The operational techniques and activities that are used to fulfil requirements of quality.
- *Quality control chart.* A graphical record of the monitoring of control samples which helps to determine the reliability of the results.
- *Quantitative analysis.* This is normally taken to mean the numerical measurement of one or more analytes to the required level of confidence.
- *Reagent.* A test substance that is added to a system to bring about a reaction or to see whether a reaction occurs (e.g. an analytical reagent).
- *Reagent blank.* A solution obtained by carrying out all steps of the analytical procedure in the absence of a sample.
- *Recovery.* The fraction of the total quantity of a substance recoverable following a chemical procedure.
- *Reference material.* This is a material or substance, one or more of whose property values are sufficiently homogeneous and well established to be used for the calibration of an apparatus, the assessment of a measurement method or for assigning values to materials.
- *Repeatability.* Precision under repeatability conditions; that is, conditions where independent test results are obtained with the same method on identical test items in the same laboratory, by the same operator, using the same equipment within short intervals of time.
- *Reproducibility.* Precision under reproducibility conditions, that is, conditions where test results are obtained with the same method on identical test items in different laboratories, with different operators, using different equipment.
- *Robustness.* For an analytical procedure, this is a measure of its capacity to remain unaffected by small, but deliberate variations in method parameters, and provides an indication of its reliability during normal usage. It is sometimes referred to as ruggedness.

- *Sample.* A portion of material selected from a larger quantity of material. The term needs to be qualified; for example, representative sample, subsample and so on.
- *Selectivity (in analysis).* Qualitative – the extent to which other substances interfere with the determination of a substance according to a given procedure. Quantitative – a term used in conjunction with another substantive (e.g. constant, coefficient, index, factor, number etc.) for the quantitative characterization of interferences.
- *Signal-to-noise ratio.* A measure of the relative influence of noise on a control signal. Usually taken as the magnitude of the signal divided by the standard deviation of the background signal.
- *Solvent extraction.* The removal of a soluble material from a solid mixture by means of a solvent or the removal of one or more components from a liquid mixture by use of a solvent with which the liquid is immiscible or nearly so.
- *Speciation.* The process of identifying and quantifying the different defined species, forms or phases present in a material or the description of the amounts and types of these species, forms or phases present.
- *Standard (all types).* A standard is an entity established by consensus and approved by a recognized body. It may refer to a material or solution (e.g. an organic compound of known purity or an aqueous solution of a metal of agreed concentration) or a document (e.g. a methodology for an analysis or a quality system). The relevant terms are as follows:
  - *Analytical standard (also known as a Standard solution).* A solution or matrix containing the analyte which will be used to check the performance of the method/instrument.
  - *Calibration standard.* The solution or matrix containing the analyte (measurand) at a known value with which to establish a corresponding response from the method/instrument.
  - *External standard.* A measurand, usually identical with the analyte, analysed separately from the sample.
  - *Internal standard.* A measurand, like but not identical with the analyte, which is combined with the sample.
  - *Standard method.* A procedure for carrying out a chemical analysis which has been documented and approved by a recognized body.
- *Standard addition.* The addition of a known amount of analyte to the sample to determine the relative response of the detector to an analyte within the sample matrix. The relative response is then used to assess the sample analyte concentration.
- *Stock solution.* This is generally a standard or reagent solution of known accepted stability, which has been prepared in relatively large amounts of which portions are used as required. Frequently, such portions are used following further dilution.

- *Sub-sample.* This may be either (i) a portion of the sample obtained by selection or division, (ii) an individual unit of the lot taken as part of the sample or (iii) the final unit of multi-stage sampling.
- *True value.* A value consistent with the definition of a given quantity.
- *Uncertainty.* A parameter associated with the result of a measurement that characterizes the dispersion of the values that could reasonably be attributed to the measurand.

1.13   Summary

The methodology for trace elemental analysis requires an understanding of a whole range of inter-related issues centred around the sample, sample preparation, analysis, data interpretation/presentation and quality assurance. This chapter has highlighted some of the most important aspects. In addition, the main strategies for calibration are discussed, including the preparation of a standard solution.

2

Sampling and Storage

LEARNING OBJECTIVES
• To be aware of the need to develop a sampling strategy. • To appreciate the different sampling regimes required for soil, water and air samples. • To highlight sample storage issues and their remedies. • To be aware of some sample preservation protocols.

2.1   Introduction

The main objective in any sampling strategy is to obtain a representative portion of the bulk sample. This requires a detailed plan of how to carry out the sampling. Therefore, the planning of the sampling strategy is an important part of the overall analytical procedure as the consequences of a poorly defined sampling strategy, as well as costing both time and money, could well lead to getting the wrong answer. An overall generic checklist for effective sampling is proposed:

- What are the overall objectives in sampling (e.g. to assess whether the land is contaminated; assess whether the table formulation contains content within its prescribed limit)?
- What will you do if these objectives are not met (i.e. re-sample or revise objectives)?
- Have arrangements been made to obtain the samples from the specific locations?
- Do you have permission to access the specific locations (e.g. the landowner in the case of a potential contaminated site)?
- Have alternative plans been prepared in case not all locations can be sampled (e.g. due to physical obstructions)?

Practical Inductively Coupled Plasma Spectrometry, Second Edition. John R. Dean.
© 2019 John Wiley & Sons Ltd. Published 2019 by John Wiley & Sons Ltd.

- Is specialized sampling equipment needed and is it available?
- Are the people involved in sampling experienced in the type of sampling required and are they available?
- Have all analytes to be determined been identified?
- Has the level of detection for each analyte been specified?
- Have methods of analysis been specified for each analyte?
- What specific quality assurance/quality control protocols are required?
- Are appropriate quality control samples available, for example, appropriate certified reference materials?
- What type of sampling approach will be used? Random, systematic, judgemental or a combination of these?
- How many samples are needed?
- How many sample locations are there?
- How many test samples are needed for each method?
- Do you need exploratory samples first?

Sampling is therefore complex and requires a clear understanding of the remit in which you will need to operate. It is often necessary to think of the answers to the questions of What? How? Why? When? to address the specific types of sample and their matrices prior to sampling. [*Note:* From a practical perspective, reasons may prevent the actual taking of a sample: e.g. the site has a concrete path or building preventing access to the underlying soil; the river may be in flood conditions, so access is considered too dangerous and there is unavailability of specialist equipment for air sampling.]

2.2   Sampling Soil

Soil is a heterogeneous material with significant variations possible within a single sampling site due to different topography, farming procedures, soil type (e.g. clay content), drainage and the underlying geology. In addition, contaminants can be distributed across a site in a random, uniform (homogenous), stratified (homogenous within sub-areas) or a gradient (Figure 2.1). However, some clues to the potential distribution of contaminants on the proposed site can be done by performing a desk top study. A desk top study includes the investigation of historic maps (to identify the location of the potential source(s) of any contaminants, e.g. buildings or storage areas) and document archives (to identify records of what was going on at the site previously, e.g. former smelting works). Based on the results of the desk top study it may then be relevant to undertake a pilot study of the site (rather than go ahead and perform a full site study). [*Note:* A pilot study seeks to reinforce the results from the desk top

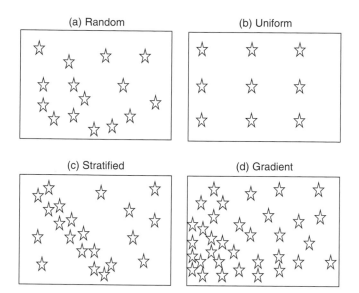

Figure 2.1 Potential contaminant distributions across a site.

study with some actual contaminant determinations as the basis of how to proceed further or not.]

Soil sampling can be done, for example, by using an auger (Figure 2.2), spade or trowel. A hand auger (e.g. corkscrew-type) allows a sample to be acquired from a reasonable depth (e.g. up to 2 m) where as a trowel is more appropriate for surface material. As all three devices are made of stainless steel the risk of contamination is reduced; however, great care needs to be taken to avoid cross-contamination from one sampling position to another. Care is needed to decontaminate (clean) the sampling device between each sample to avoid cross-contamination. Once the sample has been obtained it should be placed inside a suitable container (e.g. a geological soil bag, Kraft®), sealed and clearly labelled with a permanent marker pen. [*Note:* On the sample container, record an abbreviated sample location and/or number, date of sampling, depth sample taken from and, name of person collecting the sample.]

After obtaining the soil sample replace any unwanted soil and cover with a grass sod, if appropriate. The sample should then be transported back to the laboratory for pre-treatment. In the laboratory the soil sample (in its sample bag) should be dried either by air drying (left in a contamination-risk free area) or drying cabinet. If information is likely to be required on the original fresh-weight of sample and/or moisture content, then such details (e.g. an accurate weight) need to be obtained in relation to the sample prior to drying.

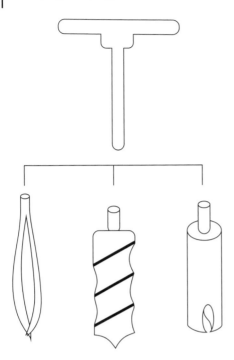

Figure 2.2 A hand-held auger (with options for three different sampling tools).

The duration of drying and temperature are variable but typically air drying at <20°C may require 7 days whereas in a drying cabinet at 40°C may be 48 hours. However, consideration needs to be given to the potential loss of contaminants due to the use of raised temperature (e.g. As and Hg). After drying and prior to sieving, it may be necessary to physically remove stones and large roots. Then, the sample should be sieved through a nylon or stainless-steel mesh. Typically, the soil samples should be sieved to <2 mm particle size. Depending on the information likely to be required, with respect to the sample result and its implication, it may be necessary to reduce the particle size of the sample still further, for example to <125 μm.

In addition, and subsequently, it may be necessary to reduce the overall quantity of the sample required for the subsequent sample treatment/analysis while still retaining the sample homogeneity. This may be done using a process called 'coning and quartering'. The process involves decanting the soil sample on to an inert and contamination-free surface, for example, a clean sheet of polythene, to form a cone. The cone is then manually divided into four quarters, using, for example, a stainless-steel trowel. Then, two opposite quarters of the cone are removed and re-formed into a new, but smaller cone. By repeating the process as many times as necessary, a suitable sized sub-sample (e.g. 5 g) can be obtained. The representative (of the whole) sub-sample is now ready for subsequent sample preparation and analysis.

2.3   Sampling Water

Water is a major constituent (approximately 70%) of the Earth's surface and is available in a variety of different types or classifications including surface waters (e.g. rivers, lakes and runoff), ground waters and spring waters, wastewaters (e.g. mine drainage, landfill leachate and industrial effluent), saline waters, estuarine waters and brines, waters resulting from atmospheric precipitation and condensation (e.g. rain, snow, fog and dew), process waters, potable (drinking) waters, glacial melt waters, steam, water for subsurface injections and water discharges (including waterborne materials and water-formed deposits).

It may appear that water is homogenous but in fact, in most cases, it is not. Spatial and temporal variation in water makes it heterogeneous; making it often difficult to obtain a representative sample. For example, spatial variation can occur within a lake due to changes in flow (i.e. a lake will often have at least one inlet source via a stream and an outlet flow via a river), differences in chemical composition (i.e. due to the different underlying geology of the lake bed) as well as temperature variation (i.e. a deep lake will be cooler than a shallow lake due to the sun's thermal heating). In addition, temporal variation, because of time differences, can occur due to heavy precipitation (e.g. snow and rainfall) as well as seasonal changes resulting in low lake water levels (e.g. leading to a concentration of contaminants) and vice versa.

Water samples are collected using the sampling device shown in Figure 2.3. It is essentially an open tube with a closure mechanism at either end; the tube is made of either stainless steel or polyvinyl chloride (PVC). Between 1 and 30 l of sample can be collected. The closed sampling device is lowered into the water to the desired depth using a distance calibrated line. Care and thought needs to be given to the potential risk of contamination from the person (and any associated peripherals) during the sampling. For example, if the sample is to be taken from a lake then the person sampling may be in a boat. Every care should be taken such that the composition of the boat does not influence the sample being taken. Then, both ends of the device are mechanically, and remotely, opened for a short time. After closing both ends the sampler is brought back to the surface and the sample transferred into a suitable container, for example, amber glass is the preferred sample container for water samples; the use of plastic containers is discouraged for water samples as they have the potential to leach metal contaminants into the acquired sample. Typically, the representative sub-sample would be stored at <4°C prior to extraction/pre-concentration and analysis.

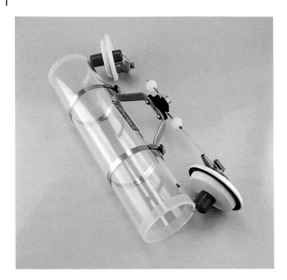

Figure 2.3 An illustration of a spring-loaded water sampling device. *Source:* Courtesy of Dynamic Aqua-Supply Ltd, Canada: http://www.dynamicaqua.com/watersamplers.html.

2.4 Sampling Air

Contaminants in the air result from both anthropogenic and natural sources. Anthropogenic sources, that is, those derived from human activity, largely occur because of burning fossil fuels and include emissions from power plants, motor vehicles, controlled burning practices (e.g. agriculture and forest management), fumes from sprays (e.g. paint) and municipal waste incineration and gas (methane) generation. Whereas natural sources include volcanic activity, wind-generated dust from exposed land and smoke from wild fires.

Air sampling can be classified into two different sample types: the sampling of air particulates on filters (passive sampling) and gaseous vapour sampling on a sorbent (non-passive sampling). While the sampling approaches are different, both types seek to determine the presence of either naturally volatile air-borne contaminants or those contaminants that become air-borne because of other activities; for example, wind generated.

In passive sampling, air-borne material diffuses on to fibre glass or cellulose fibre filters where the material is collected (Figure 2.4). The collected material is then extracted or digested from the filter prior to analysis. In non-passive sampling air-borne material is actively pumped through a sorbent (e.g. ion-exchange resin or polymeric substrate) and collected (Figure 2.5). By sampling a known quantity of air ($10\text{--}500\,\text{m}^3$) quantitative sampling is possible. After collection the sample containing sorbent tube is sealed and transported back to the laboratory for analysis. Alternatively, total suspended particulates can

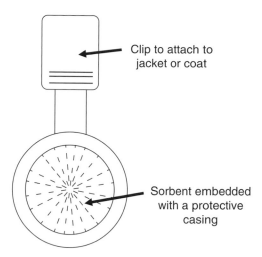

Figure 2.4 An example of a device used for the passive sampling of air. Sorbent embedded with a protective casing clip to attach to jacket or coat.

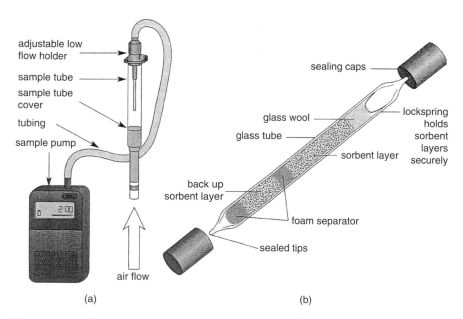

Figure 2.5 A schematic diagram of an air sampling device (a) sorbent tube sampling system and (b) a typical sorbent tube.

be sampled from the air ($150\,\mathrm{m^3\,h^{-1}}$) using a High-Volume sampler (Figure 2.6). In a 'Hi-Vol' sampler air is drawn through a quartz-microfibre filter for 24 hours. The filter can collect air-borne particulates at the impact-inlet in the

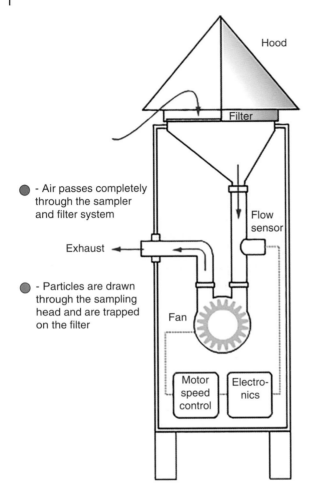

Hood

Filter

- Air passes completely through the sampler and filter system

Flow sensor

Exhaust

- Particles are drawn through the sampling head and are trapped on the filter

Fan

Motor speed control

Electro-nics

Figure 2.6 A schematic diagram of a high-volume sampler for collection of total suspended particulates. This work is licensed under the Creative Commons Attribution-ShareAlike 4.0 License. *Source:* https://creativecommons.org/licenses/by/4.0/ © CC BY 4.0 AU, Queensland Government, Australia. www.qld.gov.au/environment/pollution/monitoring/air-pollution/samplers.

particulate matter (PM) range 10 μm or 2.5 μm. The filter can then be transported back to the laboratory for subsequent analysis.

2.5 Sample Storage

The storage of samples is one of those things that you must do between sample collection, sample preparation and the subsequent analysis. Issues that can be controlled, and hence addressed, include the choice of storage vessel and its location as well as preservation techniques and the time duration prior to analysis.

The major factors affecting sample storage are as follows:

- Chemical; for example, contamination of the sample from its container, oxidation of compounds and photochemical decomposition of compounds.
- Physical; for example, sorption of metals on to container wall.
- Biological; for example, decomposition of compounds due to microorganisms.

It would be impossible to eliminate these factors, so the normal procedure is to reduce them as far as possible.

The choice of storage container for (liquid) samples is important. However, no container is inert. Ultimately, the container can influence the sample integrity in two ways: sorption of the analyte on to its surface or leaching of contaminants in to the sample. Therefore, the choice of the container composition is important (as well as its preparation prior to sample contact). Sample containers (beakers, volumetric flasks, sample jars) are made of glass (e.g. borosilicate glass) or a type of plastic (e.g. low/high density polyethylene, PE; polypropylene, PP; polyvinyl chloride, PVC; polytetrafluoroethylene, PTFE; polycarbonate, PC; perfluoroalkoxy fluorocarbons, PFA).

So, while borosilicate glass may appear to be inert it is not; its composition, primarily of, SiO_2, Al_2O_3, Na_2O, K_2O, B_2O_3, CaO and BaO as well as elemental impurities inherent within the constituents can pose problems in the laboratory. In addition, glass also presents a chemically highly reactive surface in the form of Si-OH groups that, depending on the pH of the stored solution, can act as an effective ion-exchange medium. For example, under alkali conditions the Si-OH surface will become $Si\text{-}O^-$; allowing metals ions to be retained. It is therefore normal practice when using glass in metal analysis to acidify the solution thereby ensuring that the potentially reactive glass surface is as inert as possible. This is normally achieved by addition of 1% nitric acid ('analytical reagent' grade) to the sample solution; it is important to use the highest grade of reagent (nitric acid) as possible to reduce the risk of contamination from its inherent impurities. While plastics may appear to be inert, they are not; while their composition does vary their formation, often using metal catalysts (e.g. Al, Cr, Ti, V and Zn), can result in contamination of the sample solution.

As the main vessels used for quantitative work (e.g. preparation of a series of calibration standards) are volumetric flasks and vials (e.g. sample container for an autosampler), it is also necessary to consider the importance of the stopper (glass/plastic) and cap, for volumetric flasks and vials, respectively. Contact between the sample solution and the stopper is often minimal, that is, sample container is stored vertically, except when you shake the contents. [*Note:* It is normal practice before sub-sampling the volumetric flask to invert a few times to ensure that homogeneity within the vessel is restored; it may not be visually apparent but some 'settling' may have taken place within the storage container.]

Stoppers for volumetric flasks come in a variety of materials including glass and plastic; often with a glass volumetric flask, you might have a plastic stopper. It is good practice to select a stopper that is colourless; that is, transparent; a coloured stopper, by definition, has a potential contamination risk.

Pre-cleaning of the storage container is also very important to reduce contamination prior to sample contact. The following generic procedure is recommended for glass or plastic containers (e.g. volumetric flask and beaker):

1) Wash the container in detergent to remove any previous sample residue.
2) Soak the container overnight in an acid bath (i.e. 10% v/v nitric acid).
3) Rinse with deionized, distilled water.
4) Repeat the rinse step at least twice more with deionized, distilled water. It is not necessary to fill the container to the top with deionized, distilled water; simply add approximately 10–20% water (as a percentage of the total volume of the container), agitate the water in the container, ensuring that the water is in contact with all internal surfaces; then discard.
5) Add your sample solution in deionized, distilled water into the container.

[*Note:* A 10% v/v nitric acid solution is made up from 10 ml of concentrated nitric acid in every 100 ml volume of deionized, distilled water.]

2.6   Sample Preservation

A limited range of preservation techniques are available for liquid samples but are essentially fulfilling a limited number of functions that include: controlling pH (e.g. to minimize Si-O$^-$ interactions for metal ions), temperature reduction (e.g. to reduce micro-organism activity) and removal of light (e.g. to reduce the possibility of photochemical decomposition). To reduce the potential for photo chemical and/or microorganism degradation of elemental species samples are normally stored in the refrigerator at 4°C (short-term storage) or freezer (−18°C) for longer-term storage. Storage under these conditions reduces most enzymatic and oxidative reactions; in addition, samples should be stored in amber/brown vessels. This has the additional benefit for light sensitive compounds when they are not being stored in the refrigerator/freezer. The potential risk of sample oxidation can also be minimized by the reduction of air (and hence oxygen) in the sample vessel. A reduction of air in the sample storage vessel can be achieved by ensuring the correct sized vessel is used for the sample and that the sample fills the vessel. Examples of methods of sample preservation for liquid samples are shown in Table 2.1.

A solid sample by its physical appearance and texture will have less contact points with the storage container than a liquid. Never the less the storage container needs to be selected on its relative inertness. However, often some pre-treatment needs to take place prior to sample storage. In the case of a soil

Table 2.1 Examples of sample preservation techniques for metals.

Element	Container[a]	Sample volume (ml)	Preservation method	Maximum holding time
Total (except Hg)	Glass or PVC or PP or PTFE	1000	Pre-acidified container with 5 ml of concentrated HNO_3 (high purity) to pH <2	6 months
Total Mercury (Hg)	Glass or PVC or PP or PTFE	600	Pre-acidified container with 5 ml of concentrated HNO_3 (high purity) to pH <2	28 days
Dissolved (except Hg)	Glass or PVC or PP or PTFE	1000	Filter at site using 0.45 μm filter into pre-acidified container with 5 ml of concentrated HNO_3 (high purity) to pH <2	6 months
Dissolved mercury (Hg)	Glass or PVC or PP or PTFE	1000	Filter at site using 0.45 μm filter into pre-acidified container with 5 ml of concentrated HNO_3 (high purity) to pH <2	28 days

a) PVC, polyvinyl chloride; PP, polypropylene; PTFE, polytetrafluoroethylene.

sample, for example, it is important to dry the soil sample in an oven at a temperature of 40 °C for at least 48 hours to allow moisture to be removed. The soil sample having previously been stored in a Kraft™ paper bag; this allows moisture to evaporate with time. Once dried the sample can then be stored in either a fridge (at 4°C) for a few weeks or freezer (at −18°C) for an extended period in a suitable container; for example, a glass container. Alternatively, the sample could be freeze-dried and then stored in the fridge or freezer.

For gaseous samples it is normally prudent to analyse them as soon as possible. As the concentration of gaseous samples often involves a form of trapping on a sorbent it could be envisaged that the sorbent trap could be stored for a limited period of time in the fridge at 4°C (and in the dark).

2.7 Summary

This chapter has focused on the different methods of sampling solids, liquids and gases. In each case, some context is provided to the type of samples likely to be encountered. In addition, sample storage has been considered; it is recommended that for metal analysis glass containers are normally used.

3

Sample Preparation

LEARNING OBJECTIVES

- To appreciate the different approaches available for preparation of samples for inductively coupled plasma spectrometry.
- To appreciate the range of approaches available for sample pre-treatment of aqueous samples.
- To be aware of the most important variables in liquid–liquid extraction.
- To understand the concept of pre-concentration using ion exchange chromatography.
- To be aware of the procedures for carrying out acid digestion (hot-plate and microwave) in a safe and controlled manner.
- To be aware of other decomposition methods, for example, fusion and dry ashing.
- To understand the potential for chemical speciation of mercury and arsenic species.
- To understand the relevance of selective-extraction methods for soil studies.
- To be able to carry out single extraction procedures using ethylenediamine tetraacetic acid (EDTA), acetic acid and diethylenetriamine pentaacetic acid (DTPA) in a safe and controlled manner.
- To be able to carry out a sequential extraction procedure on a soil sample in a safe and controlled manner.
- To be aware of an approach to perform a non-specific extraction (CISED).
- To be able to carry out a physiologically based extraction (*in vitro* gastrointestinal extraction) method.
- To be able to carry out a physiologically based extraction (*in vitro* inhalation bioaccessibility extraction) method.

3.1   Introduction

The ideal sample preparation for inductively coupled plasma (ICP) analysis is none! Unfortunately, this is never the case and some form of sample preparation is always required. Often, this is to enable presentation of the sample or a

Practical Inductively Coupled Plasma Spectrometry, Second Edition. John R. Dean.
© 2019 John Wiley & Sons Ltd. Published 2019 by John Wiley & Sons Ltd.

representative form of it to the sample introduction system for the ICP. For aqueous samples, often the minimum requirement is filtration to remove particulates; occasionally some form of pre-concentration is required using ion exchange, the formation of metal chelates or chromatographic separation to obtain element species specific information. For (semi-) solid samples, the normal practice is to actively digest the sample matrix to liberate the metal. This is often accomplished using a combination of acid(s) and heat (conventional or microwave heating). Alternate, but related, formats include the use of heat only (muffle furnace) with or without the addition of a modifier (fusion). Occasionally, a laser can be used to ablate a solid sample which is introduced directly into the ICP. For atmospheric samples, leaching or digestion of the collection filter may occur in a similar manner to solid samples.

3.2 Aqueous Samples

Aqueous samples come in a variety of forms including surface waters (e.g. sea, river, and lake), groundwater (e.g. natural aquifers), atmospheric precipitation (e.g. rain and snow), estuarine water (e.g. river-sea tidal areas), drinking water (e.g. reservoir and river) and waste water (e.g. run-off from a disused mine or discharge from an industrial plant). While they may all appear to be similar in nature, for example, water, their composition can vary enormously with both temporal (i.e. time) and spatial variation. Determining the metal content of waters can be challenging in terms of the sensitivity of the analytical technique, that is, inductively coupled plasma-atomic emission spectrometry (ICP–AES) or inductively coupled plasma-mass spectrometry (ICP–MS). This is because, and fortunately, the 'background' concentration levels of toxic elements in waters are low. In the case of ICP–MS it is likely that it can determine the metals, within its operating conditions, in the sample such that only filtration may be required for sample preparation; this also may be the case in some instances for ICP–AES. In this situation, the sample is filtered through a 0.2 μm filter to remove particulates which otherwise might cause blockages in connecting tubing or the nebulizer of the instrument. The basis of other sample preparation approaches for aqueous samples are therefore based on pre-concentration techniques; that is, approaches that allow a large volume of aqueous sample to be taken and its metal content concentrated to allow detection. A variety of pre-concentration approaches are available.

3.2.1 Liquid–Liquid Extraction

Liquid–liquid extraction is a traditional and robust method of sample preparation. The basis of it is that an equilibrium exists for the analyte between two immiscible phases; that is, an aqueous and an organic phase. Agitating by shaking the mixture establishes an equilibrium such that the phase containing the analyte, A (in this case a metal chelate complex) with an organic phase, such as methylisobutyl ketone (MIBK) can be expressed as:

$$A(aq) \Leftrightarrow A(org) \tag{3.1}$$

where (aq) and (org) are the aqueous and organic phases, respectively. The ratio of the activities of A in the two solvents is constant and can be represented by:

$$K_d = \{A\}_{org} / \{A\}_{aq} \tag{3.2}$$

where K_d is the distribution coefficient. While the numerical value of K_d provides a useful constant value, at a temperature, the activity coefficients are neither known or easily measured. A more useful expression is the fraction of analyte extracted (E), often expressed as a percentage:

$$E = C_o V_o / (C_o V_o + C_{aq} V_{aq}) \tag{3.3}$$

or

$$E = K_d V / (1 + K_d V) \tag{3.4}$$

where C_o and C_{aq} are the concentrations of the metal chelate in the organic phase and aqueous phases, respectively; V_o and V_{aq} are the volumes of the organic and aqueous phases, respectively, and V is the phase ratio V_o / V_{aq}.

For a one step liquid–liquid extraction, K_d must be large, that is, >10, for quantitative recovery (>99%) of the metal chelate in the organic phase, for example, MIBK. This is a consequence of the phase ratio, V, which must be maintained within a practical range of values: $0.1 < V < 10$). Typically, two or three repeat extractions are required with fresh organic solvent (i.e. MIBK) to achieve quantitative recoveries. To determine the amount of metal chelate extracted after successive multiple extractions the following is used:

$$E = 1 - \left[1 / (1 + K_d V) \right]^n \tag{3.5}$$

where n = number of extractions. For example, if the volume of the two phases are equal ($V = 1$) and $K_d = 3$ for a metal chelate, then four extractions ($n = 4$) would be required to achieve >99% recovery.

Classically, this is done using the technique of chelation extraction. The most common approach is based on the use of the following chelating (i.e. metal complexing agent) ammonium pyrrolidine dithiocarbamate (APDC) (Figure 3.1) into an organic solvent; typically, MIBK (also known as 4-methylpentan-2-one). The extraction is pH sensitive and so can be used for selective-extraction of specific elements (Table 3.1).

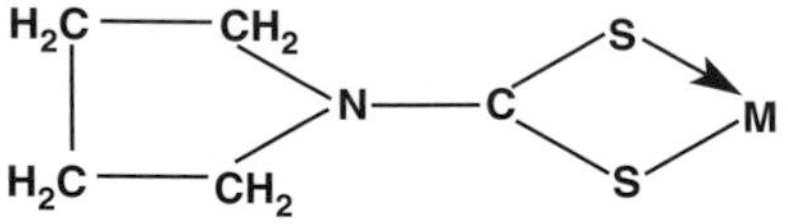

Figure 3.1 Structure of ammonium pyrrolidine dithiocarbamate (APDC). [*Note:* The ammonium ion is displaced by the metal, M.]

Table 3.1 Metal chelation extraction: pH dependence of APDC chelation.

pH range	Metals that form APDC complexes
2	W
2–4	Nb, U
2–6	As, Cr, Mo, V, Te
2–8	Sn
2–9	Sb, Se
2–14	Ag, Au, Bi, Cd, Co, Cu, Fe, Hg, Ir, Mn, Ni, Os, Pb, Pd, Pt, Ru, Rh, Tl, Zn

Source: Adapted from Kirkbright, G.F. and Sargent, M., Atomic Absorption and Fluorescence Spectroscopy, Academic Press, London, 1974.

3.2.1.1 Procedure for APDC Extraction in to MIBK

An aqueous sample (100 ml) is placed in a 200 ml extraction tube and 5 ml of APDC (1% w/v) is added. The mixture is mixed on a vortex mixer for 10 seconds. Then, the metal chelate mixture is extracted by addition of MIBK (10 ml) and further vortex mixed for 20 seconds. Finally, after allowing the mixture to stand for 10 minutes the organic layer containing the metal chelate is removed with a pipette and stored in a polyethylene bottle at 4°C for analysis. [*Note:* Further aliquots of APDC may be added to improve the efficiency of the extraction, and combined, prior to storage and analysis.] [*Note:* The metal standards for quantitation should also be prepared using the same approach; this allows for any discrepancy in extraction efficiency to be compensated for.]

Other common chelating agents used in the solvent extraction of metals are sodium diethyl-dithiocarbamate, 8-hydroxyquinoline (or oxine) and dithizone (or diphenylthiocarbazone). Sodium diethyl-dithiocarbamate is useful for extracting V, Cr, Mn, Fe, Co, Ni, Cu, Zn, Ga, Se, Nb, Mo, Ag, Cd, In, Sn, Te, W, Re, Hg, Tl, Pb, Bi, U and Pu; 8-hydroxyquinoline for Mg, Al, Ca, Sc, Ti, V, Mn, Fe, Co, Ni, Cu, Zn, Ga, Sr, Zr, Nb, Mo, Ru, Pd, Cd, In, Sn, Hf, W, Pb, Bi, Ce, Th, Pa, U and Pu and, dithizone for Mn, Fe, Co, Ni, Cu, Zn, Pd, Ag, Cd, In, Sn, Pt, Au, Hg, Tl, Pb, Bi and Po.

3.2.2 Ion Exchange

An ion exchange resin can be used to pre-concentrate metal ions from solution. If a cation-exchange resin is used it will separate metal ions (+ve charged ions) in solution. The basis of the ion exchange process is as follows: a metal ion in solution (M^{n+}) is concentrated on the resin ($_nR.SO_3^-H^+$) to form $(R.SO_3^-)_n.M^{n+}$:

$$_nR.SO_3^-H^+ + M^{n+} = \left(R.SO_3^-\right)_n.M^{n+} + H^+ \tag{3.6}$$

To desorb the retained metal ions back into solution requires the addition of acid (H^+):

$$\left(R.SO_3^-\right)_n.M^{n+} + H^+ = _nR.SO_3^-H^+ + M^{n+} \tag{3.7}$$

This approach can be used in both a batch process (i.e. in a beaker) or via an on-line flow through system directly to the nebulizer of the ICP.

A specific variation of this approach is the use of *chelation ion exchange*, which allows selectivity between monovalent ions (e.g. Na^+) and divalent ions (e.g. Pb^{2+}) by a factor of approximately 5000:1. The most common type of chelation ion exchange resin is Chelex-100; the resin contains iminodiacetic acid functional groups. The selectivity of Chelex-100, in acetate buffer at pH 5, is $Pd^{2+} > Cu^{2+} \gg Fe^{2+} > Ni^{2+} > Pb^{2+} > Mn^{2+} \gg Ca^{2+} = Mg^{2+} \ggg Na^+$. In contrast at pH 4 the selectivity is: $Hg^{2+} > Cu^{2+} > Pb^{2+} \ggg Ni^{2+} > Zn^{2+} > Cd^{2+} > Co^{2+} > Fe^{2+} > Mn^{2+} > Ca^{2+} \ggg Na^+$. Whereas, at pH 9, in the presence of 1.5 M $(NH_4)_2SO_4$, the selectivity is: $Co^{2+} > Ni^{2+} > Cd^{2+} > Cu^{2+} > Zn^{2+} > Ca^{2+} \ggg Na^+$. This approach has been applied in on-line flow through systems (see Section 4.4).

3.2.2.1   Procedure for Batch Ion Exchange Extraction

An aqueous sample (100 ml) is placed in a 200 ml extraction tube and 5 g of $AG^{\circledR}$ 50 W (strong cation-exchange resin, Bio-Rad) or Chelex 100 (chelating ion exchange resin, Bio-Rad) is added. The mixture is mixed or shaken gently for 1 hour. Then, the sample is filtered or decanted from the resin. To liberate the pre-concentrated metal from the resin acid must be added. Finally, 1% HNO_3 (v/v) is added to the resin to liberate the metal; the acid phase is removed with a pipette and stored in a polyethylene bottle at 4°C for analysis.

3.3   Solid Samples

Solid and semi-solid samples come in a variety of forms including soil (e.g. contaminated, recreational or agricultural land), metals and their alloys (e.g. steel production), particulate matter (e.g. urban dust collected from filters; i.e. airborne matter PM10 or ground level deposits), chemical (e.g. purity of chemicals and reagents), semiconductor (e.g. purity of silicon wafers), geological (e.g. mineral content, precious metal prospecting), nuclear (e.g. waste treatment), fruit and vegetables (e.g. health and regulatory limits) and sediment (e.g. river bed deposition). All are very different in terms of their matrices and normally require a decomposition approach to convert them into an aqueous sample for introduction via the nebulizer into the ICP. Often, the issues in terms of ICP–AES and ICP–MS analysis are derived from the sample matrix and method of decomposition that can lead to spectral and mass interferences.

Sample pre-treatment is often required prior to sample preparation per se. This might involve (freeze) drying of a sample (to remove moisture); sample types might include soil, biological materials, fruit and vegetables. The pre-treatment might also include grinding and sieving, of the dried sample, to reduce the particle size fraction of the sample (particularly soil samples). Grinding may take place in either a mortar and pestle or ball mill (Figure 3.2)

(a)

(b)

Figure 3.2 Preparation of a soil sample (a) grinding and sieving of a soil sample and (b) ball milling of a soil sample.

on the dried soil sample. Using a mortar and pestle (Figure 3.2a) involves grinding the dried soil sample prior to sieving through a 2 mm sieve and retaining for subsequent sample preparation. Whereas in the ball mill process (Figure 3.2b) the sample is placed in a suitable container with agate balls; when the lid and cap are placed on the container it is then subjected to vibration and rotation in the mill that results in crushing of the soil sample by the agate balls. The original sample is now ready for the next stage of the overall process.

3.3.1 Decomposition Techniques

Decomposition techniques involving acid digestion involve the use of mineral or oxidizing acids and an external heat source to decompose the sample matrix. The choice of an individual acid or combination of acids is dependent upon the nature of the matrix to be decomposed (Table 3.2). The most obvious example of this relates to the digestion of a matrix containing silica (SiO_2); for example,

Table 3.2 The common acids used for digestion of samples.

Acid(s)	Comments
Hydrochloric (HCl)	Useful for salts of carbonates, phosphates, some oxides and some sulfides. A weak reducing agent; not generally used to dissolve organic matter (boiling point: 110 °C).
Sulfuric (H_2SO_4)	Useful for releasing a volatile product; good oxidizing properties for ores, metals, alloys, oxides and hydroxides; often used in combination with HNO_3 (boiling point: 338°C). *Caution*: H_2SO_4 must never be used in PTFE vessels (PTFE has a melting point of 327°C and deforms at 260°C).
Nitric (HNO_3)	Oxidizing attack on many samples not dissolved by HCl; liberates trace elements as the soluble nitrate salt. Useful for dissolution of metals, alloys and biological samples (boiling point: 122°C).
Perchloric ($HClO_4$)	At fuming temperatures, a strong oxidizing agent for organic matter (boiling point: 203°C). *Caution*: violent, explosive reactions may occur – care is needed. Samples are normally pre-treated with HNO_3 prior to addition of $HClO_4$.
Hydrofluoric (HF)	For digestion of silica-based materials; forms SiF_6^{2-} in acid solution (boiling point: 112°C); *caution* is required in its use; glass containers should not be used, only plastic vessels. In case of spillages, calcium gluconate gel (for treatment of skin contact sites) should be available prior to usage; evacuate to hospital immediately if skin is exposed to liquid HF.
Aqua regia nitric/hydrochloric)	A 1:3v/v mixture of HNO_3: HCl is called aqua regia; forms a reactive intermediate, NOCl. Used for metals, alloys, sulfides, and other ores/best known because of its ability to dissolve Au, Pd and Pt.

a geological material. In this situation, the only appropriate acid to digest the silica is hydrofluoric acid (HF). HF is the reagent used for dissolving silica-based materials as it can convert the silicates into more volatile species in solution:

$$SiO_2 + 6HF = H_2\left(SiF_6\right) + 2H_2SO_4 \tag{3.8}$$

No other acid or combination of acids will liberate the metal of interest from the silica matrix.

Once the choice of an acid is made (Table 3.2), the sample is placed in an appropriate vessel for the decomposition stage. The choice of vessel, however, depends on the nature of the heat source to be applied. Most commonly, the acid digestion of solid matrices is carried out in open glass vessels (beakers or boiling tubes) by using a hot-plate or multiple-sample digestor (Figure 3.3).

3.3.1.1   Procedure for Acid Digestion A

Approximately 1.0 ± 0.0001 g of soil sample is accurately weighed into a digestion tube (250 ml volume). Then, 0.5–1.0 ml of water is added to form a slurry. To the slurry, 7 ml of concentrated (12.0 M) HCl is slowly added, followed by 2.3 ml of concentrated (15.8 M) HNO_3. Finally, add 15 ml of 0.5 M HNO_3 to the digestion tube. Then, allow the mixture to stand for 16 hours at room temperature. The standing of the mixture for 16 hours allows the process of oxidation of the organic matter to commence; care needs to be taken that the mixture does not get contaminated from airborne debris during this extended period. After 16 hours, raise the temperature of the reaction mixture until reflux conditions are achieved and maintain for 2 hours. Finally, allow the mixture to cool to room temperature. Quantitatively transfer the contents of the digestion tube to a 100 ml volumetric flask; quantitative transfer involves rinsing the digestion tube with additional quantities of 0.5 M HNO_3; these should also be transferred to the volumetric flask. Finally, make up to the mark with deionized, distilled water, stopper and store. Prior to analysis the mixture should be shaken and the supernatant analysed.

3.3.1.2   Procedure for Acid Digestion B

Approximately 1.0 ± 0.0001 g of soil sample is accurately weighed into a digestion tube and 10 ml of 1:1 v/v concentrated HNO_3:water added. The mixture is then heated at 95°C on a heating block for 15 minutes without boiling. After cooling at room temperature for 5 minutes, 5 ml concentrated HNO_3 is added and the sample is heated again to 95°C for 30 minutes. Then, an additional 5 ml of concentrated HNO_3 is added until no brown fumes are given off. The mixture is then allowed to evaporate to <5 ml. After cooling, 2 ml of water and 3 ml of 30% H_2O_2 are added and heated (<120°C) until effervescence subsides and the solution cools. Additional H_2O_2 can be added until effervescence ceases;

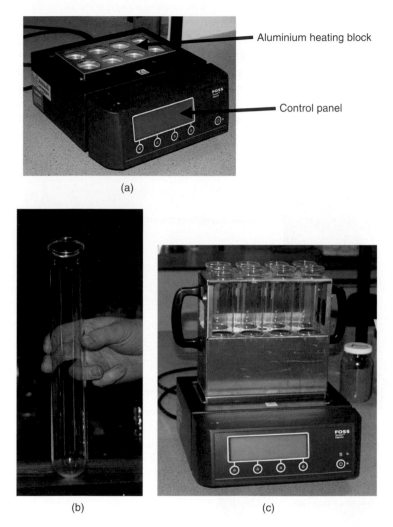

Figure 3.3 Multiple-sample digester (a) sample digester, (b) digestion tube for sample and (c) fully assembled multiple-sample digester.

however, no more than 10 ml H_2O_2 in total should be added. This stage is continued for two hours. Finally, the solution is evaporated to <5 ml. After cooling, add 10 ml of concentrated HCl and heat (<120°C) for 15 minutes. After cooling, filter the sample through a Whatman No. 41 filter paper into a 100 ml volumetric flask, and then make up to the mark with ultra-pure water, stopper and store. Prior to analysis the mixture should be shaken and the supernatant analysed. [*Note*: Unless an acid digestion procedure uses HF it will not produce a

100% efficient digestion; only HF can dissolve silicate. Therefore, the correct phrase to describe this type of digestion results in a 'pseudo-total'; however, the term 'total' is often used irrespective of acid type.]

An alternate form of heating is the use of scientific *microwave oven*. However, the heating of acids singly or in combination, under pressure, in a microwave oven requires a level of caution to be exercised in the use of such systems. So, it is essential that the user always consult manufacturer's safety and operational recommendations prior to use. A typical microwave oven system is shown in Figure 3.4.

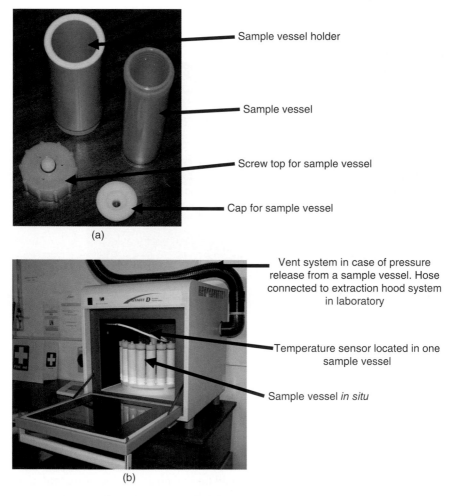

Sample vessel holder

Sample vessel

Screw top for sample vessel

Cap for sample vessel

(a)

Vent system in case of pressure release from a sample vessel. Hose connected to extraction hood system in laboratory

Temperature sensor located in one sample vessel

Sample vessel *in situ*

(b)

Figure 3.4 A typical microwave oven digestion system (a) microwave sample vessel and its component parts and (b) microwave oven (door open).

3.3.1.3   Procedure for Microwave Acid Digestion

Approximately 0.5 ± 0.0001 g of soil is accurately weighed into a 65 ml PFA (a perfluoralkoxy resin) microwave vessel pre-cleaned with concentrated acid. Then, 13 ml of aqua regia ($HCl:HNO_3$, 3:1 v/v) is carefully added into the PFA vessel and the vessel sealed with a Teflon cover. The vessel is then introduced into the safety shield of the rotor body and placed in the polypropylene rotor of the microwave. The microwave oven is operated at a temperature of 160°C, a power of 750 W, extraction time of 40 minutes and a ventilation (cooling time) of 30 minutes. After cooling, the digested sample will need to be filtered, as total digestion will not have occurred using this acid combination, into a 50 ml volumetric flask. The filtrate is diluted to the mark with deionized, distilled water prior to analysis. [*Note*: Filtering is essential to eliminate blockages; for example, the nebulizer of the ICP.]

3.3.2   Alternate Decomposition Techniques

Alternate decomposition techniques do not use acids but rely solely on heat with or without the addition of a reagent. These alternates are available in two forms, dry ashing and fusion. *Dry ashing* involves the heating of a sample in a silica or porcelain crucible in a muffle furnace at 400–800°C. The resultant residue can then be dissolved in mineral acid prior to ICP analysis. For this approach to be effective, the sample will need to contain a high proportion of combustible or organic material; for example, a high organic matter soil. While this approach will undoubtedly destroy the organic matter of the sample matrix, it can also lead to loss of some volatile elements, including Hg, Pb, Cd, Ca, As, Sb, Cr and Cu. This can be partly remedied by the addition of compounds to retard the loss of volatiles; however, its analytical use is limited due to severe disadvantages, namely losses due to volatilization, resistance to ashing by some matrices, difficulties in dissolution of some ashed material and the high risk of contamination.

3.3.2.1   Procedure for Dry Ashing

The sample, typically between 0.5–100 g (accurately weighed), is placed in an open inert vessel (e.g. silica, porcelain, or platinum crucible) and heated to 450–550°C in a muffle furnace for approximately 1–2 hours (Figure 3.5). Often the sample may be pre-charred using a Bunsen burner prior to placing in the muffle furnace. An 'ashing aid' (e.g. high purity magnesium nitrate) is often added to prevent losses. The crucible containing sample is then allowed to cool to room temperature in a desiccator and then accurately weighed. Then, the sample residue is dissolved in dilute nitric acid and transferred to a volumetric flask prior to analysis. A modified version of this approach is *sulfated ashing* in which the sample is treated with the minimum amount of concentrated sulfuric acid post-charring with a flame; the sample mixture is heated until the white dense sulfur trioxide fumes cease and prior to placing in the muffle

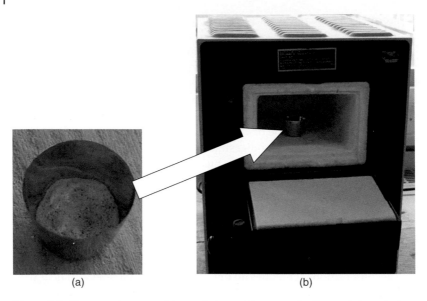

(a)   (b)

Figure 3.5 Alternate decomposition techniques (a) prepared soil sample in a platinum crucible and (b) muffle furnace.

furnace. [*Note:* Sulfated ashing may be useful when the analyte is volatile; the use of sulfuric acid acts to retain the analyte as the sulfate.]

Fusion involves the addition of an excess of reagent (10-fold) to the finely ground sample which is placed in a platinum crucible and then heated in a muffle furnace (300–1000°C). After heating for a specific period, a clear 'melt' should result, thus indicating completeness of the decomposition. After allowing to cool, the melt can be dissolved in mineral acid. Typical fusion reagents include the following:

- Sodium carbonate: 12–15 g of flux per g of sample are heated to 800°C and the resultant melt is then dissolved with HCl.
- Lithium meta- or tetraborate: a 10–20-fold excess of flux is heated to 900–1000°C and the resultant melt is then dissolved with HF.
- Potassium pyrosulfate: a 10–20-fold excess of flux is heated to 900°C and the resultant melt is then dissolved with H_2SO_4.

The addition of excess reagent (flux) can lead to a high risk of contamination, while in addition the high salt content of the final solution may lead to problems associated with ICP analysis including spectroscopic interferences as well as physical effects, such as blockages in the nebulizer of an ICP.

3.3.2.2 Procedure for Fusion

Approximately 5 ± 0.0001 g of finely ground sample is treated with 60 g of sodium carbonate (flux) and placed in a platinum crucible. The mixture is then heated in a muffle furnace (800°C) for 0.5–1 h (see Figure 3.5b). A clear 'melt' should result (indicating completeness of the decomposition). After cooling the melt is dissolved in HCl prior to analysis by ICP.

3.4   Extraction Procedures

The term extraction is specifically used instead of digestion as it refers to the removal of metals from sample matrices without destruction of the latter. This process of removing metals from sample matrices is often associated with the terms 'selective-extraction', 'physiologically based extraction' and 'speciation'. The use of *single* or *sequential methods* of extraction are required to remove metals, without altering their chemical form (speciation), from the sample matrix. This approach has been applied extensively in environmental studies; that is, soil/sediment analysis. Alternate approaches have also been applied to assesses the chemical forms of metals in contaminated and natural soils. One such approach is *the Chemometric Identification of Substrates and Element Distributions* (CISED) method. CISED is a non-specific extraction method used to assess the chemical forms of metals in contaminated soils. Other approaches have sought to mimic the environmental health risk to humans by ingestion (using *in vitro* simulated gastrointestinal extraction approaches, typified by the *Unified Bioaccessibility Method, UBM*) or inhalation (using *in vitro* simulated epithelium lung fluid extraction approaches, typified by the *Inhalation Bioaccessibility Method, IBM*).

The term *speciation* is often defined as 'the process of identifying and quantifying the different defined species, forms or phases present in a material' or 'the description of the amounts and types of these species, forms or phases present'. The reasons why speciation is important is that metals (and metalloids) can be present in many forms, for example oxidation states, or as organometallic compounds, some of which are toxic. One approach to determine the speciation of metals (and metalloids) in environmental samples has been the linking of chromatographic separation with quantitation by ICP analysis. In this situation, the use of a suitable chromatographic technique, for example, gas or liquid chromatography, is used to separate a metal complex prior to its detection by ICP analysis.

Selected examples of metal speciation include *mercury*. All forms of mercury are poisonous. However, it is methylmercury (or as the chloride, CH_3HgCl) that is the most toxic because of its ability to bioaccumulate in fish. The infamous example of the toxicity of methylmercury occurred in Minamata in Japan in 1955. It was found that methylmercury-contaminated fish consumed by pregnant women resulted in the new-born children having severe brain damage (Minamata disease). Therefore, methylmercury is routinely monitored for in fish and sediments. A variety of methods exist for the extraction, clean-up and subsequent analysis of methylmercury in samples. Most methods involve a solvent extraction of the methylmercury from the sample followed by chromatographic separation by gas chromatography (GC) or high performance liquid chromatography (HPLC). The column from the GC can be directly coupled to the ICP torch provided a heated transfer line is used to allow direct introduction of the volatile mercury species or directly by HPLC.

A limited number of certified reference materials (CRMs) are available (Table 3.3) for metal-species determination in a range of matrices including environmental, food and clinical samples. The available CRMs include the determination of particular elements (e.g. Cr[VI]), organometallic species (e.g. tributyltin) or reagent soluble species (e.g. EDTA-extractable content).

3.4.1 Procedure for Extractable Mercury from Soil and Sediment

Accurately weigh 1.0 ± 0.0001 g of homogenized soil or sediment into a microwave extraction vessel. Add 10.0 ml of 4.0 M HNO_3 to each sample. Irradiate vessels at 100°C for 10 minutes; after cooling, filter the extracts (0.22 µm glass fibre). Store extracts at ≤6°C ready for analysis (within 5 days).

3.4.2 Procedure for Speciation of Extractable Mercury from Soil and Sediment

The extract (pH 3–7) is then ready for analysis by HPLC coupled to an ICP. The individual inorganic mercury (e.g. $HgCl_2$, Hg^{2+}) and organomercurials (i.e. CH_3Hg^+ and $CH_3CH_2Hg^+$) species are separated using an isocratic (C18 column, 30 cm × 4.00 mm, 5 µm) HPLC column using a mobile phase of 30:70 methanol:water, 0.005% 2-mercaptoethanol, 0.6 M ammonium acetate, v/v.

Arsenic occurs in many chemical forms in the environment and food matrices, that is, arsenite As(III), arsenate As(V), monomethylarsonic acid (MMAA), dimethylarsinic acid (DMAA), arsenobetaine (AsB) and arsenocholine (AsC) (Table 3.4) with a range of toxicities, for example, As(III) is toxic while AsB is non-toxic. A variety of methods exist for the extraction, clean-up and subsequent analysis of arsenic species in samples. An approach for the extraction of inorganic and organoarsenicals from rice and rice-containing foods using HPLC coupled with an ICP (with mass spectrometry).

3.4.3 Procedure for Arsenic Species Extraction

An accurately weighed sample (1.0 g) is placed into a labelled 50 ml screw-cap Sarstedt tube (extraction tube) to which 10 ml of 0.28 M HNO_3 is added. Then, vortex for 10–30 seconds. After capping of the extraction tube place in a preheated block digestion system at 95°C for 90 minutes. After cooling, add approximately 6.7 g deionized water. The resulting solution was then

Table 3.3 Selected Certified Reference Materials for metal-species determination.

Area	Matrix	Identifier	Type	Metal species
Environmental	Ash, particulate and dust	BCR-545	Welding dust loaded on filter	Cr(VI)/total leachable Cr
		BCR-605	Urban dust	trimethylPb
	Soil, sediment and sludge	BCR-462	Coastal sediment	Tributyltin/dibutyltin
		ERM-CC580	Estuarine sediment	Total Hg and CH_3Hg^+
		NRCPACS-3	Harbour sediment	Tributyltin/dibutyltin
		NRCSOPH-1	Marine sediment	Tributyltin/dibutyltin
		LGC6189	River sediment	Extractable metals after 2 h reflux: As, Cd, Cr, Cu, Mn, Mo, Ni, Pb and Zn
		BCR-646	Freshwater sediment	Butyltin and phenyltin
		BCR-684	River sediment	Phosphorus: NaOH-extractable; HCl-extractable; Inorganic; Organic; and Concentrated HCl-extractable
		BCR-701	Sediment	Extractable trace elements (three step extraction): Cd, Cr, Cu, Ni, Pb and Zn
		BCR-142R	Light sandy soil	Aqua regia soluble trace elements: Cd, Ni, Pb and Zn
		BCR-700	Organic rich soil	EDTA-extractable trace elements: Cd, Cr, Cu, Ni, Pb and Zn.
				Acetic acid-extractable trace elements: Cd, Cr, Cu, Ni, Pb and Zn.

(Continued)

Table 3.3 (Continued)

Area	Matrix	Identifier	Type	Metal species
		ERM-CC018	Contaminated sandy soil	Aqua regia soluble trace elements: As, Cd, Co, Cr, Cu, Hg, Mn, Ni, Pb, V and Zn.
		ERM-CC141	Loam soil	Aqua regia soluble trace elements: As, Cd, Co, Cr, Cu, Hg, Mn, Ni, Pb and Zn.
		NIST-2701	Contaminated soil	Cr(VI)
		RTC-CRM041-30G	Soil	Cr(VI)
		RTC-CRM060-30G	Clay	Cr(VI)
		RTC-CRM061-30G	Sandy loam	Cr(VI)
		RTC-CRM206-225G	Sandy loam 3	Toxicity Characteristic Leaching Procedure (TCLP) metals: Ag, As, Ba, Cd, Cr, Hg, Pb and Se
		RTC-CRM207-225G	Loamy sand	TCLP metals: Ag, As, Ba, Cd, Cr, Hg, Pb and Se
		RTC-CRM209-225G	Sandy loam 11	TCLP metals: As, Ba, Cd, Cr and Pb
		RTC-CRM210-225G	Sandy loam 12	TCLP metals: Ag, As, Ba, Cd, Cr, Hg, Pb and Se
		RTC-CRM211-225G	Sandy loam 13	TCLP metals: Ag, As, Ba, Cd, Cr, Hg, Pb and Se
		RTC-CRM212-225G	Loamy sand 1	TCLP metals: As, Ba, Cd, Cr and Se
		RTC-CRM213-225G	Loamy sand 2	TCLP metals: Ag, As, Ba, Cd, Cr, Hg, Pb and Se
		RTC-	Sandy loam 6	TCLP metals: Extraction fluid 1 and 2: As, Ba, Cd, Cr, Hg,

	RTC-CRM217-225G	Sandy loam 8	TCLP metals: Ag, As, Ba, Cd, Cr, Hg, Pb and Se
	RTC-CRM218-225G	Loam 1	TCLP metals: Ag, As, Ba, Cd, Cr, Cu, Hg, Ni, Pb, Sb, Se, V and Zn
	LGC6181	Sewage sludge	Aqua regia soluble trace elements: As, Cd, Cr, Cu, Fe, Hg, Mn, Ni, Pb, V and Zn.
	BCR-143R	Sewage sludge	Aqua regia soluble trace elements: Cd, Cr, Mn, Ni, Pb and Zn.
	BCR-144R	Sewage sludge (domestic origin)	Aqua regia soluble trace elements: Cd, Co, Cr, Cu, Hg, Mn, Ni, Pb and Zn.
	BCR-145R	Sewage sludge (mixed origin)	Aqua regia soluble trace elements: Cr, Cu, Ni, Pb and Zn.
	BCR-146R	Sewage sludge (industrial origin)	Aqua regia soluble trace elements: Cd, Co, Cr, Cu, Hg, Mn, Ni, Pb and Zn.
	BCR-483	Sewage sludge amended soil	EDTA-extractable trace elements: Cd, Cr, Cu, Ni, Pb and Zn. Acetic acid-extractable trace elements: Cd, Cr, Cu, Ni, Pb and Zn.
	BCR-484	Sewage sludge amended (terra rossa) soil	EDTA-extractable trace elements: Cd, Cu, Ni, Pb and Zn. Acetic acid-extractable trace elements: Cd, Cu, Ni, Pb and Zn.
Food	Fish and shellfish products	BCR-463	Tuna fish	Total and methylHg
		BCR-627	Tuna fish tissue	Total As, arsenobetaine and dimethylarsinic acid

(Continued)

Table 3.3 (Continued)

Area	Matrix	Identifier	Type	Metal species
		ERM-CE464	Tuna fish	Total and methylHg
		ERM-CE477	Mussel tissue	Butyltin compounds
		NIST-1566b	Oyster tissue	Total metals and methylHg
		NIST-1947	Lake Michigan fish tissue	Total metals, organics and methylHg
		NIST-2974A	Mussel tissue	Organics plus inorganic Hg, total Hg and methylHg
		NIST-2976	Mussel tissue	Trace elements and methylHg
		NRCDORM-4	Fish protein	Trace elements and methylHg
		NRCLUTS-1	Non-defatted lobster hepatopancreas	Trace elements and methylHg
		NRCTORT-3	Lobster hepatopancreas	Trace elements, arsenobetaine and methylHg
Clinical	Blood	NIST-955C	Toxic elements in frozen caprine blood	Trace elements plus methylHg, ethylHg, Inorganic Hg and total Hg.
	Urine	NIST-2669	Frozen human urine	Arsenic species: arsenous acid (As III), arsenic acid (As V), monomethylarsonic acid (MMAA), dimethylarsinic acid (DMAA), trimethylarsine oxide (TMAO), arsenobetaine (AsB) and arsenocholine (AsC).
	Hair	IAEA-085	Human hair (powder)	Trace elements and methylHg
		IAEA-086	Human hair (powder)	Trace elements and methylHg

BCR, Bureau Communautaire de Reference; ERMs, European Reference Materials; IAEA, International Atomic Energy Authority; LGC, Laboratory of the Government Chemist; NIST, National Institute of Standards and Technology; NRC, National Research Council of Canada; RTC, RT Corp.

Table 3.4 Arsenic compounds found in environmental samples.

Compound	Formula
Arsenious acid; arsenite; As (III)[a]	$HAsO_2$
Arsenic acid; arsenate; As (V)[a]	H_3AsO_4
Monomethylarsonic acid (MMAA)[a]	$H_2(CH_3)AsO_3$
Dimethylarsinic acid (DMAA)[a]	$H(CH_3)_2AsO_2$
Arsenobetaine (AsB)	$(CH_3)_3As^+CH_2COOH$
Arsenocholine (AsC)	$(CH_3)_3As^+CH_2CH_2COOH$

a) Compounds forming gaseous species: As (III) and As (V) form AsH_3; MMAA forms
monomethylarsine, CH_3AsH_2; DMAA forms dimethylarsine, $(CH_3)_2AsH$.

centrifuged at 3000 rpm for 10 minutes. Then, filter the supernatant through a
0.45 μm nylon syringe filter. After discarding the first 1 ml, dilute 1 g of filtrate
with 2 g of freshly prepared pH adjustment solution (0.9 g of 20% ammonium
hydroxide in 100 g of HPLC mobile phase; pH 9.85 ± 0.5) and confirm the pH is
between 6 and 8.5. The sample is then ready for analysis by HPLC coupled to
an ICP. The arsenic species are separated by isocratic anion-exchange
(PRP X100 column, 4.1 × 250 mm, 10 μm) HPLC using a mobile phase of 10 mM
ammonium phosphate dibasic $((NH_4)_2HPO_4)$ at pH 8.25 ± 0.05.

3.4.4   Single Extraction Methods

In some cases, it is possible to identify, by using *single extraction methods*,
'groups' of metals without clear identification. For example, EDTA-extractable
trace metals. Single extraction methods are available that determine the fraction of the metal that is available using a non-destructive extraction protocol.
In this situation the emphasis is on assessing the potential of the metal extracting from a matrix, for example, soil or sediment. It is noted that these single
extraction methods, evolved over many years on an empirical basis, are both
element-specific and crop-specific. These approaches are often used to estimate the potential metal plant uptake from soil (Table 3.5).

A range of single extraction methods are available using for example EDTA,
acetic acid (AA), diethylenetriaminepentaacetic acid (DTPA), calcium chloride
$(CaCl_2)$, ammonium nitrate (NH_4NO_3) and sodium nitrate $(NaNO_3)$. A generic
protocol for single (and sequential) extraction methods is shown in Figure 3.6.

3.4.4.1   Procedure for EDTA Extraction

A soil sample (2 g) is weighed into a 50 ml Sarstedt extraction tube and 20 ml of
0.05 M EDTA (pH 7.0) is added. The mixture is shaken in an end-over-end
shaker at 30 rpm for 1 hour at ambient temperature (20 ± 2 °C). Then, the mixture

Table 3.5 Selective-extraction methods which are diagnostic of plant uptake.

Extractant	Element	Correlated plant content
water	Cd, Cu, Zn	Wheat, lettuce
$0.05\,mol\,l^{-1}$ EDTA[a)]	Cd, Cu, Ni, Pb, Zn	Arable crops
$0.05\,mol\,l^{-1}$ EDTA[a)]	Se, Mo	Greenhouse crops
DTPA[b)]	Cd, Cu, Fe, Mn, Ni, Zn	Beans, lettuce, maize, sorghum, wheat
2.5% v/v acetic acid	Cd, Co, Cr, Ni, Pb, Zn	Arable crops, herbage
$1\,mol\,l^{-1}$ ammonium acetate, pH 7	Mo, Ni, Pb, Zn	Herbage, oats, rice, sorghum, Swiss chard
ammonium acetate: EDTA $(0.5\,mol\,l^{-1}:0.02\,mol\,l^{-1})$	Cu, Fe, Mn, Zn	wheat
$0.05\,mol\,l^{-1}$ $CaCl_2$	Cd, Pb	vegetable
$0.1\,mol\,l^{-1}$ $NaNO_3$	Cd, Pb	vegetable
$0.05\,mol\,l^{-1}$ ammonium nitrate	Cd, Pb	vegetable

a) EDTA, ethylenediametetraacetic acid (diammonium salt).
b) DTPA = $0.005\,mol\,l^{-1}$ diethylenetriaminepentaacetic acid + $0.1\,mol\,l^{-1}$ triethanolamine + $0.01\,mol\,l^{-1}$ $CaCl_2$.

is centrifuged at 3000g for 10 minutes. Finally, the supernatant is removed with a pipette and stored in a polyethylene bottle at 4°C for analysis.

3.4.4.2   Procedure for AA Extraction
A soil sample (1 g) is weighed into a 50 ml Sarstedt extraction tube and 40 ml of 0.43 M CH_3COOH is added. The mixture is shaken in an end-over-end shaker at 30 rpm for 16 hours at ambient temperature (20 ± 2°C). Then, the mixture is centrifuged at 3000g for 10 minutes. Finally, the supernatant is removed with a pipette and stored in a polyethylene bottle at 4°C for analysis.

3.4.4.3   Procedure for Diethylenetriaminepentaacetic Acid (DTPA) Extraction
A soil sample (2 g) is weighed into a 50 ml Sarstedt extraction tube and 4 ml of 0.005 M DTPA was added. The mixture is shaken in an end-over-end shaker at 30 rpm for 2 hours at ambient temperature (20 ± 2°C). Then, the mixture is centrifuged at 3000g for 10 minutes. Finally, the supernatant is removed with a pipette and stored in a polyethylene bottle at 4°C for analysis.

3.4.4.4   Procedure for Calcium Chloride ($CaCl_2$) Extraction
A soil sample (2 g) is weighed into a 50 ml Sarstedt extraction tube and 20 ml of 0.01 M $CaCl_2$ added. The mixture is shaken in an end-over-end shaker at

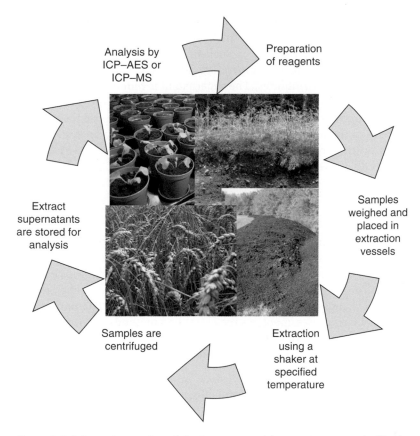

Figure 3.6 Schematic overview of single or sequential extraction protocols. (Single extraction protocols seek to mimic metal plant uptake from soil samples whereas sequential extraction protocols seek to characterize metal – mineral linkages in soils, and their release.)

30 rpm for 3 hours at ambient temperature ($20 \pm 2°C$). Then, 12 ml is decanted into a centrifuge tube and centrifuged at $3000g$ for 10 minutes. Finally, the supernatant is removed with a pipette and stored in a polyethylene bottle at 4°C for analysis.

3.4.4.5   Procedure for Ammonium Nitrate (NH_4NO_3) Extraction

A soil sample (2 g) is weighed into a 50 ml Sarstedt extraction tube and 5 ml of 1.0 M NH_4NO_3 added. The mixture is shaken in an end-over-end shaker at 50–60 rpm for 2 hours at ambient temperature ($20 \pm 2°C$). Then, the supernatant is passed through an acid-washed filter paper into a 50 ml polyethylene bottle (discard the first 5 ml of the filtrate) where it is stabilized by addition of 1 ml of concentrated HNO_3 and stored at 4°C prior to analysis. [*Note:* If solids remain, centrifuge or filter through a 0.45 μm membrane filter.]

3.4.4.6   Procedure for Sodium Nitrate (NaNO₃) Extraction

A soil sample (2 g) is weighed into a 50 ml Sarstedt extraction tube and 5 ml of 0.1 M $NaNO_3$ added. The mixture is shaken in an end-over-end shaker at 120 rpm for 2 hours at ambient temperature ($20 \pm 2°C$). Then, the mixture is centrifuged at 4000g for 10 minutes. The supernatant is removed with a syringe and filtered through a 0.45 µm membrane filter into a 50 ml polyethylene bottle. Finally, 2 ml of concentrated HNO_3 is added to a 50 ml volumetric flask, made up to volume with the filtered extract and stored at 4°C prior to analysis.

For details on the preparation of the following solutions – 0.05 M EDTA, 0.43 M CH_3COOH, 0.005 M DTPA, 0.01 M $CaCl_2$, 1.0 M NH_4NO_3 and 0.1 M $NaNO_3$ – please see Appendix 3.A.

3.4.5   Sequential Extraction

Sequential extraction procedures have been developed that allow the isolation of a metal-containing soil phase; for example, an exchangeable component. The information can then be used to identify the potential likelihood of metal release, transformation, mobility or availability as the soils are exposed to weathering, pH changes, changes in land use and their associated implications for environmental risk assessment. As the extraction methods are non-specific in nature, it has been necessary to define limits over which they can operate. The specifically defined soil phases are as follows:

- *Water-soluble, soil solution, sediment pore water.* This phase contains the most mobile and hence potentially available metals species.
- *Exchangeable species.* This phase contains weakly bound (electrostatically) metal species that can be released by ion exchange with cations such as Ca^{2+}, Mg^{2+} or NH_4^+. Ammonium acetate is the preferred extractant as the complexing power of acetate prevents re-adsorption or precipitation of released metal ions. In addition, acetic acid dissolves the exchangeable species, as well as more tightly bound exchangeable forms.
- *Organically bound.* This phase contains metals bound to the humic material of soils. Sodium hypochlorite is used to oxidize the soil organic matter and release the bound metals. An alternative approach is to oxidize the organic matter with 30% hydrogen peroxide, acidified to pH 3, followed by extraction with ammonium acetate to prevent metal ion re-adsorption or precipitation.
- *Carbonate bound.* This phase contains metals that are dissolved by sodium acetate acidified to pH 5 with acetic acid.
- *Oxides of manganese and iron.* The acidified hydroxylamine hydrochloride releases metals from the manganese oxide phase with minimal attack on the iron oxide phases. Amorphous and crystalline forms of iron oxides can be discriminated between by extracting with acid ammonium oxalate in the dark and under UV light, respectively.

To accommodate these different phases, the sequential extraction procedure consists of three (main) stages, plus a final (residual fraction) stage (Figure 3.7), as follows:

- *Step 1 'exchangeable fraction'*. The metals released in this stage are described as bioavailable and hence are the most mobile in the environment. The

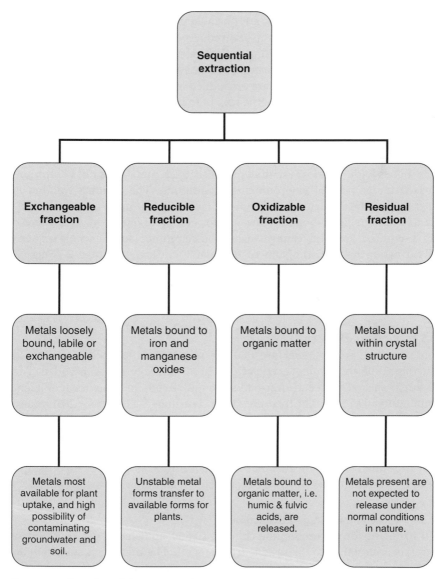

Figure 3.7 Overview of the sequential extraction method.

metals are weakly absorbed on the soil or sediment surface by relatively weak electrostatic interaction; the metals can be released by either ion exchange processes or co-precipitated with carbonates (already present in the sample). In the environment changes in the ionic composition (that could influence adsorption-desorption reactions), or lowering of the pH could cause mobilization of metals from this fraction.

- *Step 2 'reducible fraction'.* The metals released are described as bound to iron/manganese oxides. The metals are unstable under reduction conditions. In the environment any changes in the redox potential (E_h) could induce the dissolution of these oxides leading to their release from the soil or sediment.
- *Step 3 'oxidizable fraction'.* The metals released are described as those bound to organic matter. The metals are embedded in the organic composition, for example, humic/fulvic acids, of the soil. In the environment this may be considered as a stable configuration and one that is the lowest risk to humans.

It is common to analyse for trace metals in the residual fraction. In this situation, the latter should contain naturally occurring minerals which may hold trace metals within their crystalline matrices. Such metals are not likely to be released under normal environmental conditions. The residual fraction is digested by using a 'pseudo-total' approach with aqua regia as most metal pollutants are not silicate-bound. However, for complete digestion, hydrofluoric acid is required. For establishing a mass balance protocol for the whole sequential extraction method the sum of the metal in Stage 1 + Stage 2 + Stage 3 + Residual fraction should equate to the total metal content of the sample (soil or sediment). The use of this mass balance approach is a good quality assurance check on the laboratory procedures adopted.

3.4.5.1 Procedure for Stage 1 Extraction

A soil sample (1 g) is weighed into a 80–100 ml polytetrafluoroethylene (PTFE) centrifuge tube and 40 ml of acetic acid (0.11 M) (i.e. Solution A) is added. The mixture is shaken in an end-over-end shaker at 30 rpm for 16 hours at ambient temperature ($22 \pm 5°C$). The mixture is then centrifuged at 3000g for 20 minutes. Then, the supernatant is removed with a pipette and stored in a polyethylene bottle at 4°C prior to analysis.

The residue is washed with 20 ml of water by shaking for 15 minutes. Then, centrifuge the residue for 20 minutes at 3000g and discard the supernatant. Break the resultant 'cake' formed during centrifugation prior to stage 2.

3.4.5.2 Procedure for Stage 2 Extraction

Add 40 ml of hydroxylammonium chloride (0.5 M, adjusted to pH 2 with nitric acid) (i.e. Solution B) to the residue from stage 1. The mixture is shaken in an end-over-end shaker at 30 rpm for 16 hours at ambient temperature ($22 \pm 5°C$). The mixture is then centrifuged at 3000g for 20 minutes. Then, the supernatant is removed with a pipette and stored in a polyethylene bottle at 4°C prior to analysis.

The residue is washed with 20 ml of water by shaking for 15 minutes. Then, the residue is centrifuged for 20 minutes at 3000g and the supernatant discarded. Break the resultant 'cake' formed during centrifugation prior to stage 3.

3.4.5.3   Procedure for Stage 3 Extraction

Add carefully, to avoid losses due to any violent reaction, 10 ml of hydrogen peroxide (8.8 M) – (i.e. Solution C) to the residue from stage 2. Allow the sample to digest for one hour with occasional manual stirring. Ensure the container is covered with a watch glass (or similar) to prevent losses. Then, continue the digestion by heating the sample to 85 ± 2°C (with occasional manual stirring for the first 30 minutes) for 1 hour in a water bath or similar. Remove the watch glass and reduce the volume of liquid present to 2–3 ml by further heating. Then, add a further 10 ml of hydrogen peroxide (Solution C), cover with the watch glass, and heat to 85 ± 2°C for 1 hour in a water bath (with occasional manual stirring for the first 30 minutes). Again, remove the watch glass and reduce the volume of liquid to approximately 1 ml by further heating. Add 50 ml of ammonium acetate (1.0 M) (i.e. Solution D) to the cooled, moist residue. Then, shake the mixture in an end-over-end shaker at 30 rpm for 16 hours at ambient temperature (20 ± 5°C). The mixture is then centrifuged at 3000g for 20 minutes. Finally, the supernatant is removed with a pipette and stored in a polyethylene bottle at 4°C prior to analysis.

For details on the preparation of solutions A–D please see Appendix 3.B.

3.4.6   CISED

The CISED method is a non-specific extraction method used to assess the chemical forms of metals in contaminated soils. The approach uses chemometric data processing, based on multivariate self-modelling mixture resolution procedures, to determine the metal distributions in soils and sediments [1]. In order to apply the CISED approach various assumptions are made including:

- The soil/sediment under investigation consist of a mixture of discrete physico-chemical components with distinct major element compositions and that the trace elements, under investigation, are distributed amongst these components.
- The physico-chemical components will dissolve to different extents; as the reagent strength increases each solution will contain differing proportions of each of the components of the soil/sediment.
- That within any given physico-chemical component all elements are dissolved congruently.

The experimental arrangement for the extraction is shown in Figure 3.8.

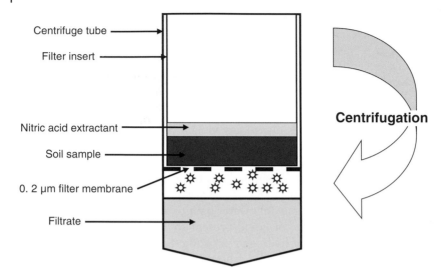

Centrifuge tube

Filter insert

Nitric acid extractant

Soil sample

0. 2 µm filter membrane

Filtrate

Centrifugation

Figure 3.8 Schematic diagram of the CISED centrifuge tube arrangement.

3.4.6.1 Procedure for CISED

A soil sample (2.0 g) is accurately weighed into a Vectaspin 20 polypropylene centrifuge tube (Whatman) with a 0.2 µm filter insert. Then, sequentially a non-specific series of reagents are added into the centrifuge tube. The order of reagents is as follows: extract 1 = distilled water; extract 2 = distilled water; extract 3 = 0.01 M HNO_3; extract 4 = 0.01 M HNO_3; extract 5 = 0.05 M HNO_3; extract 6 = 0.05 M HNO_3; extract 7 = 0.1 M HNO_3; extract 8 = 0.1 M HNO_3; extract 9 = 0.5 M HNO_3; extract 10 = 0.5 M HNO_3; extract 11 = 1.0 M HNO_3; extract 12 = 1.0 M HNO_3; extract 13 = 5.0 M HNO_3 and extract 14 = 5.0 M HNO_3. The sample-containing tube is then centrifuged for 10 minutes at 1034g and the resultant leachate is collected in a clean sample bottle for analysis. To each of the 0.1, 0.5, 1.0 and 5.0 M HNO_3 extracts was added 0.25, 0.50, 0.75 and 1.0 ml, respectively, of H_2O_2; the H_2O_2 was added to re-dissolved any precipitated organic matter. All solutions were then made to 10 ml in a volumetric flask prior to analysis.

By then applying the chemometric algorithm it is possible to estimate the physico-chemical characteristics of the sample. Cave et al. [1] compared their CISED generated data with the determined mineralogy for NIST SRM 2710 (a highly contaminated soil) and were able to identify nine chemically distinct soil components: pore water residual solutes; organic; easily exchangeable; a Cu-Zn dominated phase; a Pb-dominated phase; amorphous Fe oxide/oxyhydroxide; crystalline Fe oxide; Fe-Ti oxide and Mn-Fe-Zn oxide.

3.4.7 *In Vitro* Gastrointestinal Extraction Method

In contrast to the single or sequential extraction methods an alternate procedure has been developed that seeks to mimic the environmental health risk to humans by ingestion. Ingestion of an environmental sample, for example soil, can occur unintentionally because of hand-to-mouth contact. That is, that an individual has been in direct contact with the soil and some of the material has been retained on the palm of the hand or under their finger nails. By periodic placing of the hands in their own mouth or nail biting direct soil ingestion occurs. [*Note:* Obviously the thorough washing of hands and scrubbing under nails after contact with soil would result in a considerably lower (or nil) risk to the individual.] Intentional consumption of soil is also known to occur in certain civilized cultures. The so-called geophagy behaviour of individuals is done because of religious belief or for medicinal purposes (e.g. the relief of morning sickness). [*Note:* the type of material consumed in the case of geophagy is carefully controlled by the individual; on that basis the individual sources and selects a certain type of material, for example Calabash chalk (Figure 3.9).]

Several ingestion methods have been developed and applied. This makes the language of the different approaches more complicated than may ultimately be necessary. In the literature the methods are described under the following names (and others as well): physiologically based extraction test (PBET); *in vitro* gastrointestinal extraction or UBM. In each case the different methods are broadly similar but can differ in the composition of chemicals used and the type of stages applied. Ultimately, each is seeking to describe the human body

Figure 3.9 A photograph of Calabash Chalk. (On the left-hand side, unrefined, while on the right-hand side, shaped.)

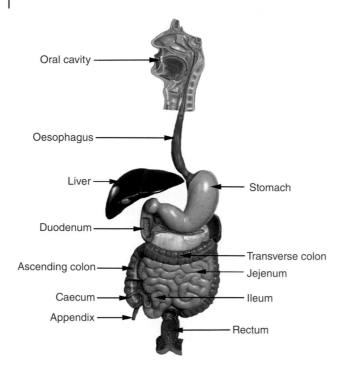

Oral cavity

Oesophagus

Liver

Stomach

Duodenum

Ascending colon

Transverse colon

Jejenum

Caecum

Ileum

Appendix

Rectum

Figure 3.10 Schematic layout of the human gastric-intestinal system.

processes that occur when the sample is introduced into the mouth on route to the stomach and potential absorption in the intestines prior to excretion (Figure 3.10). The gastrointestinal tract is extremely complex and hence the different methods all make assumptions about chemical composition, pH variation, different 'holding' times, agitation, and anaerobic/aerobic conditions. Details for the preparation of simulated (saliva, gastric, duodenal and bile) fluids are given in Appendix 3.C. The procedure outlined in the following is based on the UBM approach.

3.4.7.1 Procedure for Gastric Extraction

An accurately weighed sample (0.6 g) is placed into a 50 ml screw-cap Sarstedt tube and treated with 9 ml of simulated saliva fluid. With the screw-cap closed the soil–fluid mixture is manually shaken. Then, after 5–15 minutes, 13.5 ml of simulated gastric fluid is added. The mixture is then shaken on an end-over-end shaker maintained at $37 \pm 2°C$. After 1 hour, the pH of the soil suspensions is checked; the pH should be within the range 1.2–1.7. Then, the solution is centrifuged at 3000 rpm for 5 minutes and a 1.0 ml aliquot of supernatant is removed. To the supernatant 9.0 ml of 0.1 M HNO_3 is added. The sample is then stored at <4°C prior to analysis.

3.4.7.2   Procedure for Gastric + Intestinal Extraction

An accurately weighed sample (0.6 g) is placed into a 50 ml screw-cap Sarstedt tube and treated with 9 ml of simulated saliva fluid. With the screw-cap closed the soil–fluid mixture is manually shaken. Then, after 5–15 minutes 13.5 ml of simulated gastric fluid is added. The mixture is then shaken on an end-over-end shaker maintained at 37 ± 2°C. After 1 hour, the pH of the soil suspensions is checked; the pH should be within the range 1.2–1.7. Then, 27.0 ml of simulated duodenal fluid and 9.0 ml of simulated bile fluid are added. With the screw-cap closed the soil–fluid mixture is manually shaken. The pH of the resultant suspension is adjusted to 6.3 ± 0.5, by the drop-wise addition of 37% HCl, 1 M or 10 M NaOH, as required. The mixture is then shaken on an end-over-end shaker maintained at 37 ± 2°C for 4 hours. The soil suspension is removed. Then, the pH of the soil suspension is measured (and record); the pH should be within 6.3 ± 0.5. The soil suspension is then centrifuged at 3000 rpm for 5 minutes and a 1.0 ml aliquot of supernatant is removed. To the supernatant 9.0 ml of 0.1 M HNO_3 is added. The sample is then stored at <4°C prior to analysis.

The percentage gastric or gastrointestinal bioaccessible fraction (%GBAF or %GIBAF) can be calculated as follows:

$$\%GBAF = \left(C_{Gastric} / C_{total} \right) \times 100 \qquad (3.9)$$

$$\%GIBAF = \left(C_{GI} / C_{total} \right) \times 100 \qquad (3.10)$$

where $C_{Gastric}$ is the gastric bioaccessible concentration of metal and C_{total} is the total concentration of metal determined by aqua regia digestion (Eq. 3.9); and C_{GI} is the gastric + intestinal bioaccessible concentration of metal and C_{total} is the total concentration of metal determined by aqua regia digestion (Eq. 3.10). For each digestion, reagent blanks were also prepared. The filtrate obtained from the digestion was refrigerated (<4°C) prior to analysis.

3.4.8   *In Vitro* Simulated Epithelium Lung Fluid Method

An alternate approach that seeks to mimic the environmental health risk to humans by inhalation has been developed. Inhalation of small particulates can cause significant health problems affecting the lungs including asthma. Current environmental concerns relate to atmospheric conditions, including smog and the occurrence of small particulates, in the air of urban cities. It is recognized that an inhaled particle may reside in one of at least two 'compartments'; the extracellular environment typified by lung fluid of neutral pH and the more acidic environment within macrophages (Figure 3.11). The pioneer synthetic lung fluid (SLF), commonly referred to as Gamble's solution, has been widely used to simulate the extracellular environment of the deep lung to assess the

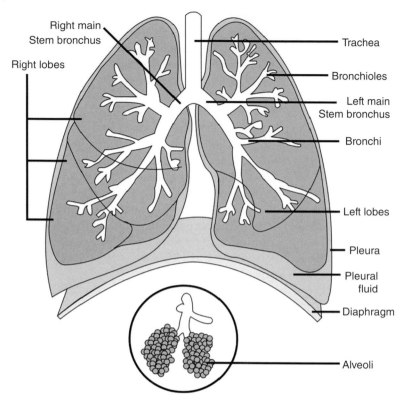

Figure 3.11 Schematic layout of the human lung system. *Source:* Reproduced with permission of Humanbodyanatomy.co. https://humanbodyanatomy.co/human-lungs-diagram/human-lungs-diagram-pictures-human-lungs-diagram-labeled-human-anatomy-diagram.

exposure of humans to inhalable pollutants. A recent addition to the different approaches available is an *in vitro* simulated epithelial lung fluid (SELF) that represents the extracellular environment of the lung. The SELF has then been applied to perform an IBM on the PM_{10} fraction of dust and soils and their related aerosols. This IBM seeks to mimic exposure from PM_{10} particles deposited in the upper respiratory tract/tracheobronchial region of our air pathway.

The published literature indicates that any SELF consists of inorganic salts and organic constituents (including surfactant lipids, large-molecular-mass proteins, low molecular-mass antioxidant proteins and organic acids). Several *in vitro* extraction methods, often based on the original 'Gamble's solution', are available; one such approach is presented here. Details for the preparation of SELF are given in Appendices (Appendix 3.D).

3.4.8.1 Procedure for Inhalation Bioaccessibility Extraction

The SELF was heated to $37 \pm 2°C$ in a thermostatically controlled water bath before use. Approximately 0.3 g of test sample was accurately weighed into a labelled 50 ml centrifuge tube to which 20 ml of the prepared SELF was added. The mixture was then shaken, on an end-over-end rotator, maintained at $37°C \pm 0.2$ for 96 hours. The resulting solution was then centrifuged at 3000 rpm for 10 minutes. Then, 1 ml of the supernatant was pipetted into a 10 ml Sarstedt tube previously holding 9 ml of 0.1 M HNO_3 and 30 µl of internal standard. The sample is then stored at <4°C prior to analysis.

The resultant residual fraction was microwave acid digested (temperature 160°C; power 750 W; extraction time 40 minutes and cooling time 30 minutes) using aqua regia ($HCl:HNO_3$ in the ratio 3:1 v/v) and stored in the fridge (4°C) for subsequent analysis.

The percentage inhalation bioaccessible fraction (%IBAF) can be calculated as follows:

$$\%IBAF = \left(C_{ibio} / C_{total} \right) \times 100 \tag{3.11}$$

where C_{ibio} is the inhalation bioaccessible concentration of metal and C_{total} is the total concentration of metal determined by aqua regia digestion. For each digestion, reagent blanks were also prepared. The filtrate obtained from the digestion was refrigerated (<4°C) prior to analysis.

3.5 Summary

This chapter looks at a whole range of procedures that can be applied to determine the total metal content of a sample as well as the procedures to obtain other important environmental aspects of the metal and its matrix. Specifically, the use of single, sequential, *in vitro* gastrointestinal and *in vitro* inhalation extraction approaches. Practical details on the use of each of these procedures are provided.

Reference

1 Cave, M.R., Milodowski, A.E., and Friel, E.N. (2004). *Geochem. Explor. Environ. Anal.* 4: 71–86.

3.A Extraction Reagents for Single Extraction Methods

3.A.1 Preparation of EDTA, 50 mM

In a fume cupboard, 146 ± 0.05 g of EDTA (free acid) is added to 800 ± 20 ml of distilled water (i.e. water that has a resistivity of $18.2 \, M\Omega \times cm$). To aid the dissolution of the EDTA stir in 130 ± 5 ml of saturated ammonia solution (prepared by bubbling ammonia gas into distilled water). Then, continue to add the ammonia solution until all the EDTA has dissolved. The resultant solution should be filtered, if necessary, through a filter paper of porosity 1.4–2.0 μm into a pre-cleaned 10 l polyethylene bottle and then diluted to 9.0 ± 0.5 l with distilled water. Then, the pH is adjusted to 7.00 ± 0.05 by the addition of a few drops of either ammonia or concentrated hydrochloric acid, as appropriate. Finally, the solution should be made up with distilled water to obtain an EDTA solution of 50 mM. [*Note:* For every freshly prepared EDTA solution analyse a sample to assess its metal impurity.]

3.A.2 Preparation of AA, 0.43 M

In a fume cupboard 250 ± 2 ml of glacial acetic acid (AnalaR or similar high purity grade) is added to approximately 5 l of distilled water (i.e. water that has a resistivity of $18.2 \, M\Omega \times cm$) in a pre-cleaned 10 l polyethylene bottle and made up with distilled water. [*Note:* For every freshly prepared acetic acid solution analyse a sample to assess its metal impurity.]

3.A.3 Preparation of Diethylenetriaminepentaacetic Acid (DTPA), 5 mM

In a fume cupboard, 149.2 g of triethanolamine (0.01 M), 19.67 g DTPA (5 mM) and 14.7 g calcium chloride are dissolved in approximately 200 ml distilled water (i.e. water that has a resistivity of $18.2 \, M\Omega \times cm$). After dissolution the solution is diluted to 9 l in distilled water. The pH is adjusted to 7.3 ± 0.5 with concentrated HCl while stirring and then diluted to 10 l in distilled water.

3.A.4 Preparation of Ammonium Nitrate (NH_4NO_3), 1 M

In a fume cupboard 80.04 g of NH_4NO_3 is dissolved in water (i.e. water that has a resistivity of $18.2 \, M\Omega \times cm$) and then made up to 1 l with water.

3.A.5 Preparation of Calcium Chloride ($CaCl_2$), 0.01 M

In a fume cupboard, 1.470 g of $CaCl_2.2H_2O$ is dissolved in water (i.e. water that has a resistivity of $18.2 \, M\Omega \times cm$) and then made up to 1 l with water. [*Note:* Verify that the Ca concentration is $400 \pm 10 \, mg \, l^{-1}$ by EDTA titration.]

3.A.6   Preparation of Sodium Nitrate (NaNO₃), 0.1 M

In a fume cupboard, 8.50 g of $NaNO_3$ is dissolved in water (i.e. water that has a resistivity of 18.2 MΩ × cm), and then made up to 1 l with water.

[*Note:* For every freshly prepared new reagent solution analyse a sample to assess its metal impurity; any detectable metal content should be considered when data is being blank subtracted.]

3.B   Extraction Reagents for the Sequential Extraction Method

3.B.1   Preparation of Solution A (Acetic Acid, 0.11 M)

In a fume cupboard, 25 ± 0.1 ml of glacial acetic acid is added to approximately 0.5 l of water (i.e. water that has a resistivity of 18.2 MΩ × cm) in a 1 l polyethylene bottle and then made up to 1 l with water. Then 250 ml of this solution (acetic acid 0.43 M) is taken and diluted to 1 l with water to obtain an acetic acid solution of 0.11 M.

3.B.2   Preparation of Solution B (Hydroxylamine Hydrochloride or Hydroxyammonium Chloride, 0.5 M)

A 34.75 g amount of hydroxylamine hydrochloride is dissolved in 400 ml of water (i.e. water that has a resistivity of 18.2 MΩ × cm). The solution is transferred to a 1 l volumetric flask and 25 ml of 2 M HNO_3 (prepared by weighing from a concentration solution) is added (the pH should be 1.5). Then, the solution should be made up to 1 l with water. [*Note:* This solution should be prepared on the same day as the extraction is carried out.]

3.B.3   Preparation of Solution C (Hydrogen Peroxide [300 mg g^{-1}], 8.8 M)

Use the H_2O_2 as supplied by the manufacturer; that is, acid stabilized to pH 2–3.

3.B.4   Preparation of Solution D (Ammonium Acetate, 1 M)

A 77.08 g amount of ammonium acetate is dissolved in 800 ml of water (i.e. water that has a resistivity of 18.2 MΩ × cm). The pH is adjusted to 2 ± 0.1 with concentrated HNO_3 and made up to 1 l with water.

[*Note:* For every freshly prepared new reagent solution analyse a sample to assess its metal impurity; any detectable metal content should be taken into account when data is being blank subtracted.]

3.C   Extraction Reagents for the Unified Bioaccessibility Method

3.C.1   Preparation of Simulated Saliva Fluid

145 mg of α-amylase (bacillus species), 50.0 mg mucin and 15.0 mg uric acid are added to a 2 l high density polyethylene (HDPE) screw top bottle. Then separately, 896 mg of KCl, 888 mg NaH_2PO_4, 200 mg KSCN, 570 mg Na_2SO_4, 298 mg NaCl and 1.80 ml of 1.0 M HCl are added into a 500 ml volume container and made up to the mark with water (*inorganic saliva component*). Into a second 500 ml volume container, 200 mg urea is added and made up to the mark with water (*organic saliva component*). Then, simultaneously the 500 ml of inorganic and 500 ml of organic saliva components are poured in to the 2 l HDPE screw top bottle. Shake the entire contents of the screw top bottle thoroughly. Then, measure the pH of this solution (*simulated saliva fluid*). The pH should be within the range 6.5 ± 0.5. If necessary, adjust the pH by adding either 1.0 M NaOH or 37% HCl.

3.C.2   Preparation of Simulated Gastric Fluid

A 1000 mg amount of bovine serum albumin, 3000 mg mucin and 1000 mg pepsin are added to a 2 l HDPE screw top bottle. Then, separately, 824 mg of KCl, 266 mg NaH_2PO_4, 400 mg $CaCl_2$, 306 mg NH_4Cl, 2752 mg NaCl and 8.30 ml of 37% HCl are added into a 500 ml volume container and made up to the mark with water (*inorganic gastric component*). Into a second 500 ml volume container is added 650 mg glucose, 20.0 mg glucuronic acid, 85.0 mg urea and 330 mg glucosamine hydrochloride and made up to the mark with water (*organic gastric component*). Then, simultaneously the 500 ml of inorganic and 500 ml of organic components are poured in to the 2 l HDPE screw top bottle. Shake the entire contents of the screw top bottle thoroughly. Measure the pH of this solution (*simulated gastric fluid*). The pH should be within the range 0.9–1.0. If necessary, adjust the pH to this range (0.9–1.0) by adding either 1.0 M NaOH or 37% HCl. Check that the combination of mixed saliva fluid (1 ml) and gastric fluid (1.5 ml) is in the pH 1.2–1.4. If the combined mixture is not within this range it is necessary to re-adjust the pH of the gastric fluid by adding either 1.0 M NaOH or 37% HCl. Re-check that the combination of mixed saliva fluid (1 ml) and gastric fluid (1.5 ml) is in the pH 1.2–1.4.

3.C.3   Preparation of Simulated Duodenal Fluid

The following are added to a 2 l HDPE screw top bottle: 200 mg of $CaCl_2$, 1000 mg bovine serum albumin, 3000 mg pancreatin and 500 mg lipase. Then, separately, 564 mg of KCl, 80 mg KH_2PO_4, 50.0 mg $MgCl_2$, 5607 mg $NaHCO_3$, 7012 mg NaCl and 180 µl of 37% HCl are added into a 500 ml volume container and made up to the mark with water (*inorganic duodenal component*). Into a second 500 ml volume container 100 mg urea is added and made up to the mark with water (*organic duodenal component*). Simultaneously, the 500 ml of inorganic and 500 ml of organic duodenal components are poured in to the 2l HDPE screw top bottle. Shake the entire contents of the screw top bottle thoroughly. Measure the pH of this solution (*simulated duodenal fluid*). The pH should be within the range 7.4 ± 0.2. If necessary, adjust the pH of the duodenal fluid by adding either 1.0 M NaOH or 37% HCl.

3.C.4   Preparation of Simulated Bile Fluid

A 222 mg amount of $CaCl_2$, 1800 mg bovine serum albumin and 6000 mg bile are added to a 2l HDPE screw top bottle. Then, separately, 376 mg of KCl, 5785 mg $NaHCO_3$, 5259 mg NaCl and 180 µl of 37% HCl are added in to a 500 ml volume container and make up to the mark with water (*inorganic bile component*). Into a second 500 ml volume container, 250 mg urea is added and made up to the mark with water (*organic bile component*). Then, simultaneously the 500 ml of inorganic and 500 ml of organic bile components are poured in to the 2l HDPE screw top bottle. Shake the entire contents of the screw top bottle thoroughly. Allow the solution to stand for approximately 1 hour, at room temperature to allow for complete dissolution of solid reagents. Measure the pH of this solution (*simulated bile fluid*). The pH should be within the range of 8.0 ± 0.2. If necessary, adjust the pH of the duodenal fluid by adding either 1.0 M NaOH or 37% HCl. Check that the combination of saliva fluid (1.0 ml), gastric fluid (1.5 ml), 3.0 ml duodenal fluid and 1.0 ml bile fluid is in the pH 6.3 ± 0.5. If the combined mixture is not within this range it is necessary to re-adjust the pH of the duodenal fluid by adding either 1.0 M NaOH or 37% HCl. Re-check that the combination of saliva fluid (1.0 ml), gastric fluid (1.5 ml), 3.0 ml duodenal fluid and 1.0 ml bile fluid is in the pH 6.3 ± 0.5.

3.D   Extraction Reagents for the *In Vitro* SELF

3.D.1   Preparation of SELF

Chemical components needed to prepare 1000 ml of SELF were prepared in two phases (inorganic and organic). To prepare the *inorganic phase* (500 ml): 6020 mg NaCl, 256 mg $CaCl_2$, 150 mg $NaHPO_4$, 2700 mg $NaHCO_3$, 298 mg KCl, 200 mg $MgCl_2$ and 72 mg Na_2SO_4 were accurately weighed into a 500 ml HDPE screw top bottle and made up to the set volume with ultra-pure water and then thoroughly mixed. To prepare the *organic phase* (500 ml):18 mg ascorbic acid,

16 mg uric acid and 30 mg glutathione were also accurately weighed into a 500 ml HDPE screw top bottle and made up to the set volume with ultra-pure water; the resulting solution was then thoroughly mixed. The *inorganic* and *organic* components were simultaneously poured into a 2 l HDPE screw top bottle containing additional constituents; 260 mg albumin, 122 mg cysteine, 100 mg dipalmitoylphosphatidylcholine (DPPC), 376 mg glycine and 500 mg mucin, this was thoroughly mixed until all the components dissolved. The pH was measured and adjusted to 7.4 ± 0.2 by adding 0.2 ml HCl.

4

Sample Introduction

LEARNING OBJECTIVES

- To be aware of the different forms of sample introduction device used for inductively coupled plasmas.
- To understand the principle of operation of a range of nebulizers.
- To appreciate the importance of a spray chamber and its use with a nebulizer.
- To be aware of the scope and operation of discrete sample introduction devices used for ICPs.
- To be aware of the scope and operation of continuous sample introduction devices used for ICPs.
- To understand the principle of coupling high performance liquid chromatography and gas chromatography systems to ICPs.
- To appreciate the advantages and limitations offered by hydride generation and cold-vapour techniques for ICPs.

4.1   Introduction

The efficient introduction of a sample into an inductively coupled plasma (ICP) is crucial for the detection of elements of different concentrations. However, equally important for a modern ICP instrument is the use of a computer-controlled autosampler in which the prepared samples can be pre-loaded for analysis (Figure 4.1). The autosampler, after programming, allows samples and standard solutions to be introduced into the ICP sequentially with sequenced rinse and wash cycles.

The solutions from the autosampler then become part of the liquid sample introduction system for the ICP. For liquid samples, the combination of a nebulizer and spray chamber is the most common approach but results in only a small portion (often <2%) of the sample reaching the ICP. While other

Practical Inductively Coupled Plasma Spectrometry, Second Edition. John R. Dean.
© 2019 John Wiley & Sons Ltd. Published 2019 by John Wiley & Sons Ltd.

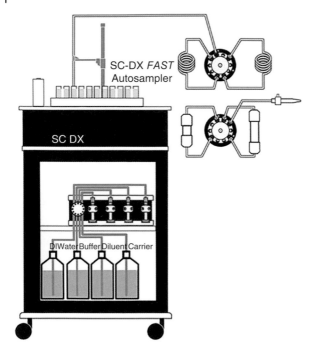

Figure 4.1 Schematic diagram for an autosampler sample presentation unit for ICP technology. *Source:* Courtesy of Elemental Scientific, Nebraska, USA. http://www.icpms .com/products/brinefast-S4.php.

approaches for the introduction of solid and liquid samples are possible, the additional complexity associated with their use has often precluded their wide-spread acceptance for routine analysis. This chapter considers the different approaches possible for the introduction of both liquid and solid samples into an ICP.

4.2 Nebulizers

A wide range of nebulizers are used for sample introduction into an ICP (Figure 4.2). As the nebulizer is the most common approach to introduce a liquid sample into the ICP, much effort has gone into the design and effectiveness of this physically small component. The following generic types of nebulizers will be considered:

- Pneumatic concentric glass nebulizer
- Cross-flow nebulizer
- V-groove (high solids) nebulizer
- Ultrasonic nebulizer

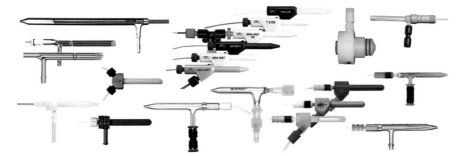

Figure 4.2 Selected common commercially available nebulizers. *Source:* https://creativecommons.org/licenses/by/4.0/.

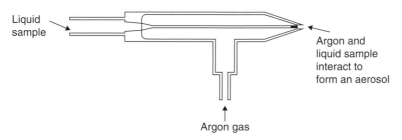

Figure 4.3 Schematic diagram of the pneumatic concentric nebulizer.

The most common nebulizer in use today is the *pneumatic concentric* glass nebulizer (Figure 4.3). The design of the nebulizer is based on the work of Gouy [1] but was first fabricated in its current format in 1979 by Meinhard [2]. It operates by using the Venturi effect. The Venturi effect, named after the Italian physicist Giovanni Battista Venturi, is the reduction in fluid pressure that results when a fluid flows through a constricted section of a pipe; the fluid in this case is argon gas. In this type of nebulizer, argon gas introduced in the side arm can exit at the nozzle, so causing the reduction in pressure. The reduced pressure results in the liquid sample being drawn up through the capillary tube and exiting through the nozzle. The physical interaction of argon gas and liquid sample causes a coarse aerosol to be produced. The coarse aerosol generated by the nebulizer cannot be introduced directly into the ICP as it would lead to either extinguishing of the plasma or to severe spectral/mass interferences due to the lowering in temperature of the plasma. Therefore, the coarse aerosol is then subjected to further treatment in a spray chamber.

In *cross-flow* nebulizers, the liquid sample and argon gas interact at perpendicular to one another. The original cross-flow nebulizer consists of two capillary needles positioned at 90° to each other with their tips not quite touching (Figure 4.4). The argon carrier gas flows through one capillary

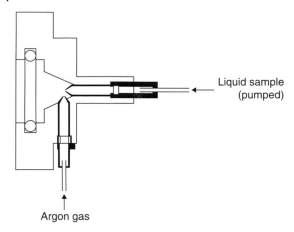

Liquid sample
(pumped)

Argon gas

Figure 4.4 Schematic diagram of the cross-flow nebulizer.

tube, while through the other capillary the liquid sample is pumped. At the exit point, the force of the escaping carrier gas is sufficient to shatter the sample into a coarse aerosol. Modifications in this design have been used for sample solutions with higher dissolved solids content; for example, the *V-groove* or Babington-type nebulizer. This type of nebulizer is designed to generate coarse aerosols from aqueous samples with high solids contents (up to 20%). The design of this nebulizer (Figure 4.5) allows the sample

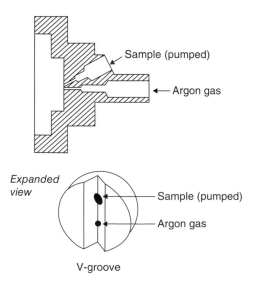

Sample (pumped)

Argon gas

Expanded view

Sample (pumped)

Argon gas

V-groove

Figure 4.5 Schematic diagram of the V-groove, high solids nebulizer.

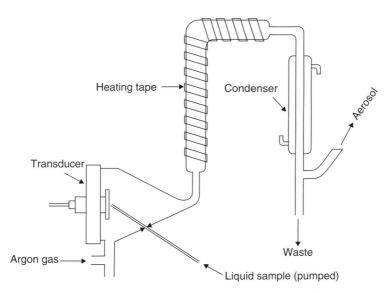

Figure 4.6 Schematic diagram of the ultrasonic nebulizer with desolvation.

solution to be pumped along a V-grooved channel. Midway along the channel is a small orifice through which the carrier gas can escape. As the sample passes over the orifice, the escaping argon gas causes generation of a coarse aerosol.

In *ultrasonic* nebulizers, the sample solution is pumped onto a vibrating piezoelectric transducer, operating at between 200 kHz and 10 MHz. The action of the vibrating crystal is enough to transform the liquid sample into an aerosol. The latter is then transported by the argon carrier gas through a heated tube and then a condenser (Figure 4.6). This has the effect of removing the solvent. Under such conditions, the aerosol is desolvated and reaches the plasma source in a fine, dry state. The major advantage of the ultrasonic nebulizer is its increased transport efficiency; of the order of 10% (compared with the pneumatic nebulizer). For further information on the fundamentals of nebulizers and their use in ICP spectrometry the reader is directed to a review published in 2014 and references therein [3, 4].

4.3 Spray Chambers and Desolvation Systems

While the ICP is a robust ionization source its water-loading capacity, and equivalent vapour flux, is known to reach a maximum at $20\,\mu l\,min^{-1}$ [5]. Therefore, introduction of the nebulizer-generated coarse aerosols directly into the plasma source would extinguish or induce cooling of the plasma, hence leading to severe matrix interferences. The inclusion of a spray chamber can result in the production of a more appropriate aerosol for the plasma source. This is because the spray chamber can:

- reduce the amount of aerosol reaching the plasma;
- decrease the turbulence associated with the nebulization process;
- reduce the aerosol particle size.

It has been determined that the ideal particle size for the plasma processes to occur, that is, atomization, followed by either excitation/emission or ionization, is around 10 μm. Several spray chamber designs are available, including the following:

- double-pass or Scott-type
- cyclonic
- single-pass, direct or cylindrical type

An ideal spray chamber should have all the following features:

- a large surface area to induce collisions and fragmentation of the coarse aerosol
- minimal dead volume to prevent dilution of the sample
- easy removal of condensed sample to waste without inducing pressure pulsing

The most common type of spray chamber is the *double-pass*. This is comprised of two concentric tubes, an inlet for the nebulizer, an exit for the finer aerosol and a waste drain (Figure 4.7). The double-pass spray chamber is positioned to allow excess liquid (aerosol condensation) to flow to waste. The nebulizer-generated aerosol is introduced into the inner tube and exits having reversed its direction (180°) into the ICP torch. Interaction of the coarse

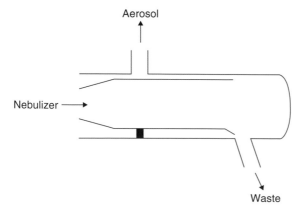

Figure 4.7 Schematic diagram of a double-pass spray chamber (Scott-type).

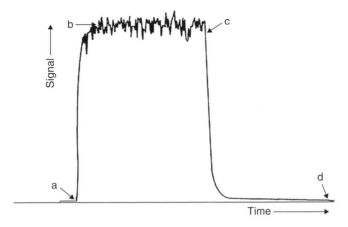

Figure 4.8 A typical time–signal profile for ICP analysis: (a) sample introduced via nebulizer/ spray chamber; (b) start of steady-state signal; (c) loss of steady-state signal commences and (d) signal decreases to 1% of the steady-state signal.

aerosol with the internal surfaces of the double-pass spray chamber leads to the production of a finer aerosol (with consequent excess liquid going to waste). This design of the spray chamber also acts to reduce the turbulence of the nebulizer-generated aerosol, hence leading ultimately to greater signal stability. A drawback of this design of spray chamber is the presence of so-called 'dead volumes' that are not easily reached by the argon nebulizer gas. This leads to an increased 'wash-out' time, that is, the time taken to remove all existing sample prior to the introduction of the next sample. Wash-out time is normally measured as the time taken for the ICP system to measure 1% of the steady-state signal (Figure 4.8).

In the *cyclonic* spray chamber (Figure 4.9), the aerosol is introduced tangentially to induce swirling. The initial process involves the aerosol swirling downwards close to the spray chamber wall. At the bottom of the spray chamber, a second inner spiral carries the aerosol to the exit point. The combined process of induced tangential flow and subsequent aerosol swirling leads to a reduction in aerosol particle size. An example of a pneumatic concentric nebulizer/cyclonic spray chamber is shown in Figure 4.10. The *single-pass*, direct or cylindrical spray chamber (Figure 4.11) often contains an impact bead for aerosol production. It is recommended that this type of spray chamber is connected to a desolvation system to reduce sample load into the plasma. *Desolvation systems* are designed to reduce the solvent, that is, water or volatile solvents, loading into the plasma. In its simplest form, it can consist of a spray chamber fitted with a thermostated water jacket.

An interesting development in the introduction of liquid samples into plasmas has been the use of *micro-nebulizers*. These devices have been developed

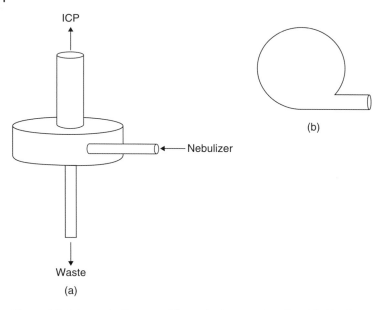

Figure 4.9 Schematic diagram of the cyclonic spray chamber: (a) side view and (b) top view.

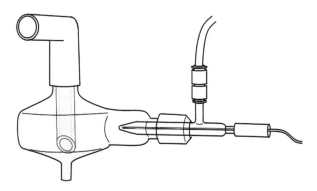

Figure 4.10 A schematic diagram of a pneumatic concentric nebulizer – cyclonic spray chamber arrangement. *Source:* Reproduced with permission of ThermoFisher.com. https:// www.thermofisher.com/de/en/home/industrial/spectroscopy-elemental-isotope-analysis/ spectroscopy-elemental-isotope-analysis-learning-center/trace-elemental-analysis-tea-information/inductively-coupled-plasma-mass-spectrometry-icp-ms-information/ icp-ms-sample-preparation.html.

for cases where sample is limited, in, for example, clinical samples. The most common approach to this has been to reduce the nebulizer gas flow rate from the conventional $1\,ml\,min^{-1}$ to the $\mu l\,min^{-1}$ level. Various pneumatic concentric micro-nebulizers have been developed and include the high-efficiency

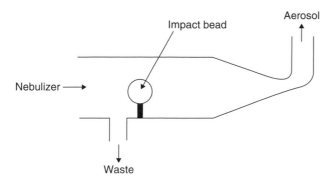

Figure 4.11 Schematic diagram of a single-pass spray chamber (direct or cylindrical type).

nebulizer and direct-injection nebulizer. An image of a complete nebulizer/ spray chamber arrangement *in situ* is shown in Figure 4.12. For further information on the fundamentals of spray chambers, the reader is directed to a comprehensive review published in 1988 on their use in ICP spectrometry [6].

4.4 Discrete Sample Introduction

The direct introduction of a sample into a plasma source can be achieved by using either the argon carrier gas, a liquid carrier stream or vapour generation. Discrete sample introduction has the advantage of presenting the plasma source with all the analyte in a short time. Therefore, while the residence time of the analyte within the plasma is short, all the analyte is available for analysis, thus leading to improved analyte sensitivity.

Laser ablation (LA) allows a solid sample to be mobilized (ablated) and transported direct to the plasma torch via the argon carrier gas. LA has the following benefits:

- applicable to any solid sample;
- no sample size requirements;
- no sample preparation, no reagents or solution waste;
- spatial characterization information available.

However, it does have some limitations, as follows:

- amount of sample ablated is dependent upon laser and sample properties;
- 'matrix-matched' standards required for calibration;
- fractionation occurs for samples containing low-melting-point elements.

One of the most common lasers used for ablation of samples is the Neodymium Yttrium Aluminium Garnet, referred to as the Nd:YAG, which

Figure 4.12 A complete nebulizer/spray chamber arrangement *in situ* (for ICP–MS) with peristaltic pump.

operates in the near infrared at 1064 nm. With optical frequency doubling, tripling, quadrupling and quintupling, the Nd:YAG laser can be operated at the following wavelengths 532, 355, 266 and 213 nm, respectively. In addition, *excimer* lasers have been used. As the operating wavelengths are dependent upon the gas, the choice of output wavelengths are 308, 248, 193 and 157 nm for XeCl, KrF, ArF and F2, respectively.

A typical experimental layout for LA coupled to a plasma source consists of an ablation chamber, a lens to allow focusing, an adjustable platform for positioning in the x-, y- and z-directions, and a charge-coupled device (CCD) camera for remote viewing of the sample surface (Figure 4.13). The sample is placed inside the ablation chamber, which is fitted with a fused silica window. The chamber is then flushed with argon carrier gas to transport the ablated sample material directly to the ICP torch (Figure 4.14). However, sample losses due occur because of:

- 'hot' sample ejected from the surface being cooled by the argon carrier gas, so leading to redeposition and
- ejected material can be lost en route to the plasma source by deposition in the connecting tubing.

Therefore, the transport process is not 100% efficient. The typical size of the crater generated when using a Nd:YAG laser is 10–100 μm in diameter. Ablation of a hard material, for example stainless steel, results in a crater profile with upraised walls (Figure 4.15a) while the ablation of a softer material, for example a semiconductor, will undergo melting and subsequent flowing away from the hot central position (Figure 4.15b).

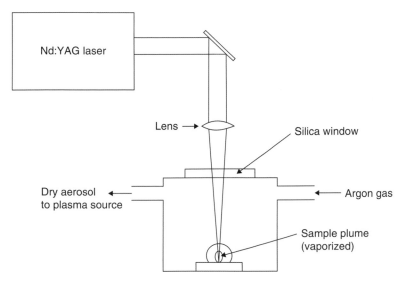

Figure 4.13 Schematic representation of laser ablation.

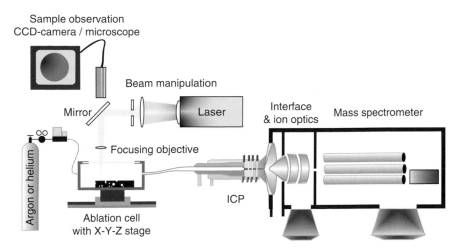

Figure 4.14 Schematic diagram of a LA–ICP system. [*Note*: With MS detection, but AES detection is also possible.] *Source:* Reproduced with permission of Elsevier.

A major disadvantage of LA is the difficulty in obtaining samples for instrument calibration. However, three different calibration strategies exist, as follows:

- matrix-matched direct solid ablation;
- dual introduction (sample-standard);
- direct liquid ablation.

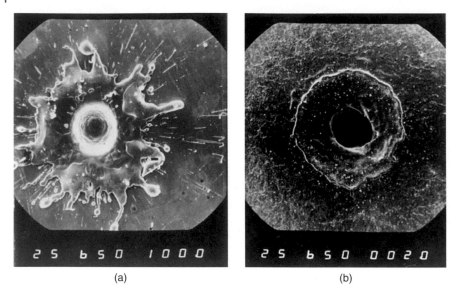

Figure 4.15 Effect of a Nd:YAG laser on the surfaces of (a) stainless steel and (b) a semiconductor material (cadmium mercury telluride).

Matrix-matched direct solid ablation using external calibration is the most common approach for LA–ICP analysis. It is essential to use matrix-matched standards as the quantity of material ablated per laser pulse is sample matrix-dependant (see the differences between hard and soft materials). The availability of certified reference materials (CRMs) allows matrix-matched standards to be used. CRMs (see Section 1.9) exist for a range of solid matrices, including steels, alloys, glass and ceramics. Several approaches are possible to produce matrix-matched standards, including the 'bricketing' of powdered material, that is, compression using a hydraulic press in the presence of a binder, addition of standard solutions to a powdered matrix, co-precipitation of the element of interest and fusion techniques (see Section 3.3.2).

In a dual introduction (sample-standard) system, laser-ablated material and solution-nebulized standards are both introduced into the ICP in a sequential manner. This allows signals from the ablated material to be compared directly with solution standards introduced via a nebulizer/spray chamber arrangement. While this approach is an attractive alternative, particularly when CRMs are difficult to obtain or simply not available, it does suffer from a major disadvantage. The drawback is the different analyte responses from the two methods due to the introduction of a 'dry' sample from LA and a 'wet' aerosol from the nebulization process. For example, in inductively coupled plasma–mass spectrometry (ICP–MS) the use of LA and hence a 'dry' sample introduction system leads to reduced oxide interferences due to the elimination of water from the sample. This disadvantage can be remedied, in part, using a desolvation system attached to the nebulizer/spray chamber.

An alternative calibration strategy uses direct liquid ablation to introduce aqueous standards into the plasma. To improve the optical absorption characteristics of the standard solution, it may be necessary to add a *chromophore*. The presence of a chromophore may lead to improved laser energy coupling to the aqueous solution standard in such a manner as to allow the ablation process to occur within the surface layers of the liquid, thereby leading to the production of a fine aerosol. The ideal characteristics of a chromophore include the following:

- an ability to absorb strongly at the laser wavelength
- does not lead to precipitation in the standard solution
- is non-toxic.

The major advantage of this approach is that it offers a new surface for each subsequent laser pulse. For qualitative work, the limited sampling size can be very useful for characterizing impurities in manufactured goods or for mineral identification in geological samples.

4.5 Continuous Sample Introduction

Flow-injection (FI) and *high performance liquid chromatography (HPLC)* are both methods of introducing aqueous samples in a flowing stream into the plasma source. FI is a 'multi-method' system that allows the user to create unique sample presentation facilities for the plasma source. In principle, any FI system will consist of a peristaltic pump, injection valve, 'sample alteration/ modification' component and an interface to the nebulizer. The mode of 'sample alteration/modification' is diverse and can be developed according to the application. For example, a FI system could be used to calibrate the instrument. By using a carrier stream in which the sample is continuously flowing discrete standard solutions could be introduced. This would provide a method for generating a standard additions calibration plot 'on-line'.

Typically, however, the 'sample alteration/modification' module may consist of (i) a low-pressure chromatography column used for retention of the analyte in preference to the sample matrix (Figure 4.16a), (ii) an 'on-line' solvent extraction system (Figure 4.16b) or (iii) a means of delivering a small discrete sample with the minimum of dilution to the nebulizer/spray chamber (Figure 4.16c).

HPLC can be useful for two reasons, namely (i) elemental species information may be obtained, so-called speciation studies (see Section 3.3) and (ii) separation of potential matrix interferences. The type of chromatography used largely depends on the nature of the analyte to be separated but includes ion-exchange, reversed-phase and size-exclusion chromatography.

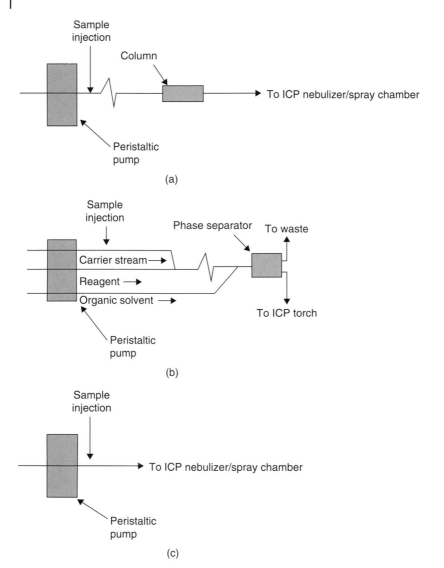

Figure 4.16 Components of a 'sample alteration/modification' module: (a) a low-pressure chromatography column, used for retention of the analyte in preference to the sample matrix; (b) an 'on-line' solvent extraction system and (c) a means of delivering a small discrete sample with the minimum of dilution to the nebulizer/spray chamber.

Interfacing an HPLC system with an ICP normally offers few challenges with respect to physical coupling of the two techniques. Liquid chromatography flow rates are often of the order of $1\,ml\,min^{-1}$, which is directly compatible with the typical aspiration rates of a conventional nebulizer/spray chamber system for ICP analysis (Figure 4.17). Therefore, the simple connection of the output from the HPLC system to the input of the nebulizer via the minimum amount of polytetrafluoroethylene (PTFE) or poly(ether ether ketone) (PEEK) tubing is all that is required, with the overall process being limited to the typical inefficiencies of the nebulizer/spray chamber system (1–2% transport efficiency). However, while the coupling offers few challenges the influence on the plasma source can be significant. The introduction of high concentrations of salts (from a mobile phase buffer) or organic solvent (from the mobile phase, e.g. methanol or acetonitrile in reversed-phase (RP) HPLC) can significantly influence the plasma temperature and its stability, as well as the sample introduction processes linked to aerosol generation and analyte transport. In addition, for ICP–MS carbon build-up on the sampling cones and lens system can lead to signal suppression and alternate isobaric interferences. This has led to instrument modifications in terms of, for example, use of reduced spray chamber temperatures ($<0°C$), as well as a new low-flow micro-concentric nebulizer coupled with low volume spray chambers, both of which act to reduce the carbon loading into the plasma.

In vapour-phase sample introduction two formats are commonly used based on *Gas Chromatography (GC)* and *Hydride Generation*. As for HPLC, GC is used to separate compounds prior to analysis. In the case of GC, the samples are separated by heat and, therefore, separation is largely based on the compounds' boiling points. The interfacing of a GC system with an ICP does not present any difficult challenges. As GC separates volatile compounds, it is essential that the interface maintains the analytes in the vapour state. This is often achieved via a heated transfer line (250°C) that allows the carrier gas (argon, helium or nitrogen) and volatile compounds to be directly transported to the ICP torch (no nebulizer/spray chamber required). This not only allows a much higher transport efficiency than in HPLC but leads to enhanced sensitivities of the elements being investigated.

For analysis of samples by GC, the analytes of interest should be volatile, thermally stable at the operating temperatures of the injection port and column oven and should give a good peak shape. However, it is possible to analyse analytes that do not meet this criteria by carrying out an additional step known as derivatization. Derivatization is carried out to modify the functionality of an analyte to facilitate separation by GC and is generally used with analytes of low volatility and those that are thermally labile, that is, compounds that could often be analysed by HPLC. Derivatization is normally done to improve the column resolution, reduce peak tailing and improve column efficiency of polar compounds; to analyse relatively non-volatile compounds, for example, those compounds with higher molecular weight, and to improve the thermal stability of some compounds. Different derivatizing reagents are available for different

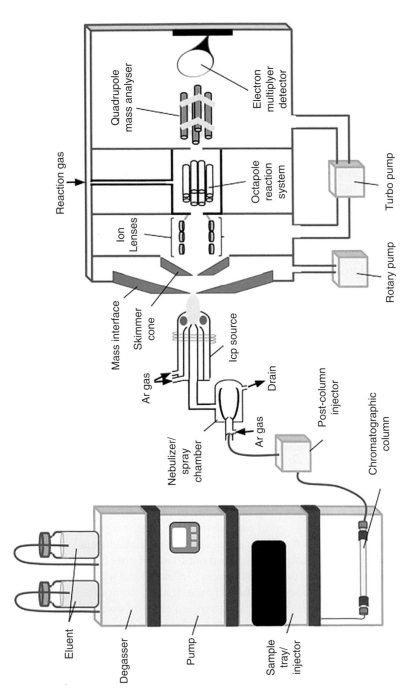

Figure 4.17 A schematic diagram of an HPLC–ICP system. [*Note*: With MS detection, but AES detection is also possible.] *Source*: Reproduced with permission of Elsevier.

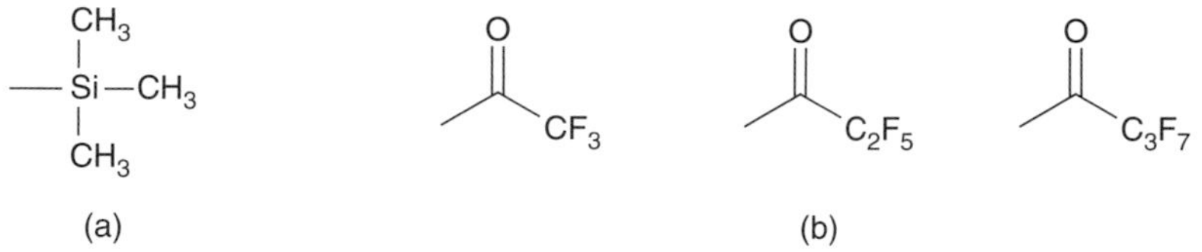

(a)   (b)

Figure 4.18 Typical derivatization reagents used for GC (a) silylation using the TMS group and (b) acylation using the acyl group.

analytes; the choice of reagent and its suitability is dependent upon the analytes functional groups. However, silylation and acylation are the two main derivatizing reagents.

Silylation involves the addition of a silyl group into the compound, often by substitution of an active hydrogen. Therefore, the addition of the silyl group reduces the polarity of the compound as well as reducing opportunities for hydrogen bonding. The resultant derivatized product is therefore more volatile and more thermally stable. Typically, silylation is done using the trimethyl-silyl (TMS) group (Figure 4.18a). Addition of the TMS group is accomplished by use of specific silylating reagents; these include N, O- bistrimethylsilyl-acetamide (BSA), N,O-bis-trimethylsilyl-trifluoroacetamide (BSTFA), N-methyl-N-trimethylsilyl-trifluoroacetamide (MSTFA) and N-trimethylsilylylimidazole (TMSI).

As with silylation, acylation produces a resultant molecule that is more volatile and less polar than the underivatized or parent analyte. This process of acylation is affected by the reaction with acyl derivatives or acid anhydrides (Figure 4.18b). Typical acid anhydride acylating agents include trifluoroacetic acid (TFAA), pentafluoropropionic anhydride (PFPA), heptafluorobutyric anhydride (HFBA) and heptafluorobutrylimidazole (HFBI). Acylating reagents are very good at reacting with highly polar functional groups that contain active hydrogens, for example, –OH, –SH and –NH, converting them into esters, thioesters and amines, respectively. As in the silylation process the resultant derivatized molecule is also less thermally labile, which results in better resolution of analyte peaks.

4.6 Hydride and Cold-Vapour Generation Techniques

Hydride generation is a vapour-phase sample introduction technique that allows the determination of ultra-trace and trace levels of analytes to be measured. It is limited to several elements that can form volatile hydrides at ambient temperature, for example, arsenic, antimony, bismuth, selenium, tellurium and tin. Under acid conditions, and in the presence of a reducing agent, for example sodium tetraborohydride, covalent hydrides are formed; for example, AsH_3, SbH_3, BiH_3, H_2Se, H_2Te and SnH_4. The principle of hydride generation can be described in four steps, as follows:

- chemical generation of the hydrides;
- collection and pre-concentration of the evolved hydrides (if necessary);
- transport of the hydrides and gaseous by-products to the ICP;
- atomization of the hydrides followed by either atomization/emission spectrometry in atomic emission spectrometry (AES) or ionization in MS.

As an example, Eq. (4.1) describes the chemical generation of the arsine hydride (AsH_3):

$$3BH_4^- + 3H^+ + 4H_3AsO_3 \rightarrow 3H_3BO_3 + 4\mathbf{AsH_3} + 3H_2O \tag{4.1}$$

However, when the basic borohydride is added to an acidic solution, excess hydrogen is liberated:

$$BH_4^- + 3H_2O + H^+ \rightarrow H_3BO_3 + 4H_2 \tag{4.2}$$

Various gas–liquid separation devices have been used for continuous, flow-injection or batch-mode separation. An example of such a separation device is shown in Figure 4.19. Cold-vapour generation is exclusively reserved for the element mercury. In this situation, the mercury present in the sample is reduced, usually with tin(II) chloride, to elemental mercury:

$$Sn^{2+} + Hg^{2+} \rightarrow Sn^{4+} + Hg^\circ \tag{4.3}$$

The mercury vapour generated is then transported to the ICP by argon carrier gas.

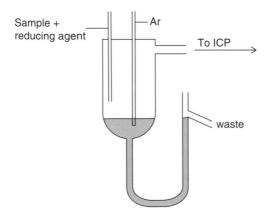

Figure 4.19 An example of a gas–liquid separation device for hydride generation.

4.7 Summary

Sample introduction is key to the efficiency of atomization/ionization in an ICP. The main approach to sample introduction is via a nebulizer/spray chamber arrangement. The range of possibilities in terms of choice of nebulizer and design of the spray chamber are discussed in this chapter. Alternative sample introduction techniques are highlighted and described – these include LA, hydride and cold-vapour generation and the use of flow-injection technologies.

References

1 Gouy, M. (1879). *Ann. Chim. Phys.* XVIII: 5–101.
2 Meinhard, J.E. (1979). Pneumatic nebulizers, present and future. In: *Applications of Plasma Emission Spectrochemistry* (ed. R.M. Barnes), 1–14. Philadelphia, PA, USA: Heyden & Son, Inc.
3 Bings, N.H., Orlandini von Niessen, J.O., and Schaper, J.N. (2014). *Spectrochim. Acta* 100: 14–37.
4 Sharp, B.L. (1988). *J. Anal. At. Spectrom.* 3: 613–652.
5 Turner, I. and Montaser, A. (1998). *Inductively Coupled Plasma Mass Spectrometry* (ed. A. Montaser). New York: VCH.
6 Sharp, B.L. (1988). *J. Anal. At. Spectrom.* 3: 939–963.

5

The Inductively Coupled Plasma

LEARNING OBJECTIVES

- To be aware of the historical development of the induction coupled plasma (ICP) as a source for atomic emission spectrometry.
- To be aware of the function of the radiofrequency generator.
- To understand and explain the principle of operation of an ICP source.
- To appreciate the concept of temperature in relation to a plasma.
- To appreciate and understand the reasons why the ICP is viewed axially or laterally.
- To understand the basic processes that occur within the ICP when a sample is introduced.
- To be aware of the necessary signal processing and instrument control required for operation of the instrument.

5.1   Introduction

While the first description of an induction coupled plasma (ICP) being generated in a Leyden jar through a coil surrounding a tube filled with rarefied air can be traced back to 1884 [1], this is concerned with the application of an ICP as an analytical technique to quantify trace elements in a range of matrices. The concept of using the ICP as a source for atomic emission spectroscopy (AES) has been jointly attributed to the independent work of Greenfield et al. [2] in the UK and Wendt and Fassel [3] in the USA. While these early instruments would not be recognizable in comparison to the modern bench top-based, PC-controlled systems the concept of using an ICP for AES as an elemental technique was born over 55 years ago. The development by commercial instrument suppliers would today be classed as rather slow, as the first complete instruments did not become available until the mid-1970s (some 10 years after its first application).

Practical Inductively Coupled Plasma Spectrometry, Second Edition. John R. Dean.
© 2019 John Wiley & Sons Ltd. Published 2019 by John Wiley & Sons Ltd.

At this time developments were taking place that would lead eventually to the coupling of an ICP to a mass spectrometer. Initial work first reported in 1974 [4] focused on the coupling of a capillary arc to a mass spectrometer. However, Gray and co-workers soon noticed that the use of a capillary arc was a limited ion source and so it was replaced by the ICP. The coupling of an ICP with a mass spectrometer was first reported in 1980 involving collaboration between Gray (University of Surrey) and Fassel at the Ames Laboratory, USA [5]. Commercialization of ICP–MS occurred much faster with the first instrument being available from Vacuum Generators (VGs) in the UK in 1984, quickly followed by a further instrument from Sciex in Canada.

A brief history of the developments of spectroscopy and spectrometry as related to the ICP for AES and mass spectrometry (MS) are presented in Table 5.1. In addition, the current commercially available instrumentation is listed (Table 5.2).

5.2 Radiofrequency Generators

The radiofrequency (RF) generator is the component that provides the power for the generation and sustainment of the plasma. The power normally ranges between 1000 and 1500 W, and is transferred to the plasma gas through a load coil or plate. The load coil acts as an antenna to transfer the RF power to the plasma. Most RF generators for ICP operate at either 27 or 40 MHz, this is to avoid interference with normal communication channels. There are two generic types of RF generator: crystal-controlled and free-running generators; the latter is the most commonly used.

5.3 Inductively Coupled Plasma Formation and Operation

A plasma is defined as the co-existence, in a confined space, of the positive ions, electrons and neutral species of an inert gas; in this case, argon. The most common plasma source used is the (radiofrequency) ICP. A plasma can be characterized by its temperature, which is often quoted as 7000–10 000 K. This vagueness in temperature is because even though the plasma is electrically neutral, it is not in thermodynamic equilibrium. Hence, it is not possible to characterize a single temperature. Four temperatures can be used to characterize the plasma; that is, excitation, ionization, electron and gas temperatures. The *excitation temperature* is a measure of the population density of energy levels (see Section 6.1.2), while the *ionization temperature* represents the population density of different ionization states, the *electron temperature*

Table 5.1 A brief history of atomic spectrometry (as related to ICP–AES and ICP–MS) [6–14].

Date	Event
1666	Isaac Newton shows that white light, from the Sun, could be dispersed into a continuous series of colours.
1678	Christian Huygens proposes the wave theory of light.
1752	Thomas Melville observes that a flame emits a bright yellow light when burning a mixture of alcohol and sea salt.
1760	Johann Henrich Lambert publishes the 'Law of Absorption'.
1776	Alessandro Volta uses static electric charges to spark various materials, noting different colours. He uses this approach to identify different gases by their emitted colours.
1786	David Rittenhouse produces the first diffraction grating.
1802	William Hyde Wollaston is the first to observe dark lines in the spectrum of the sun.
1814	Joseph von Frauenhofer invents the transmission diffraction grating.
1826	William Henry Fox Talbot observes that different salts produce colours when placed in a flame.
1851	M.A. Masson produces the first spark-emission spectroscope.
1852	August Beer publishes his paper showing that the amount of light absorbed is proportional to the amount of solute in aqueous solutions.
1859	Gustav Robert Kirchoff and Robert Wilhelm Eberhard von Bunsen discover that spectral lines are unique to each element.
1865	Julius Plucker and Johann Wilhelm published a paper on gas discharge tubes.
1868	Anders Jonas Angstrom publishes a detailed study of the wavelengths of solar spectral lines.
1869	Anders Jonas Angstrom produces the first reflection grating.
1873	Joseph Norman Lockyer observes that the emission from a high-voltage spark-induced plasma is a function of the emitting element. He proposes improved quantitation is possible by comparing the analyte emission with that from another element (internal standardization).
1877	L.P. Gouy introduces the pneumatic nebulizer for transferring liquids into a flame.
1882	Henry A. Rowland produces an improved (curved) diffraction grating.
1891	N. Tesla reports studies on electrodeless discharges.
1891	J.J. Thomson reports the first study on an inductively coupled discharge; it was used to excite various gases for observation by a direct-vision spectroscope.

(Continued)

Table 5.1 (Continued)

Date	Event
1897	Joseph Thomson discovers the electron by analysing the deflection of cathode rays in an electric field to determine its mass/charge ratio.
1898	W. Wien discovers the proton by analysing positive anode rays in an electric field to measure the mass of the hydrogen atom.
1900	Max Planck introduces the quantum concept.
1900	Frank Twyman (Adam Hilger Ltd) produces the first commercially available quartz prism spectrograph.
1905	Albert Einstein explains the photoelectric effect.
1911	First time-of-flight spectrometer built by Hammer.
1913	Niels Bohr presents his theory of the atom.
1918	A.J. Dempster constructed the first focusing magnetic mass spectrometer.
1919	F.W. Aston built his first mass spectrograph.
1920	Adam Hilger Ltd produce the first evacuated spectrograph for the determination of S (180.7 nm) and P (178.2) in steel.
1921	A.J. Dempster used a 180° magnetic sector field mass spectrometer with an electrometer. He determined the isotopic abundances for the stable Mg isotopes.
1927	Erwin Schrodinger develops his uncertainty principle, which explains the natural linewidth of spectral lines.
1928	Irving Langmuir introduces the term 'plasma'.
1929	H. Lundegardh advance flame emission spectroscopy by using an air-acetylene flame with a pneumatic nebulizer for sample introduction and a spray chamber for sample conditioning.
1930	Walther Gerlach and Eugen Schweitzer develop the concept of internal standard and the method of intensity ratios.
1934	J. Mattauch and R. Herzog develop theory for double focusing MS.
1935	A.J. Dempster built the first double focusing MS.
1936	Thanheiser and Heyes use photocells to measure intensities.
1937	Maurice Hasler (Applied Research Laboratories, ARL) introduce the first commercial grating spectrograph.
1939	George R. Harrison publishes the MIT wavelength tables.
1940	Photomultiplier tube is developed.
1940	Nier's first precision (60°) mass spectrometer for investigation of isotopic abundances and isotopic ratios.
1941	G.I. Babat reports the use of high frequency inductively coupled plasma for industrial applications.
1944	R.W. Wood produces blazed gratings.

Date	Event
1946	First time-of-flight mass spectrometer (Stephens).
1947	G.I. Babat described the process of maintaining an induction plasma on a stream of gas.
1947–1948	First commercially available 'direct-readers' optical emission spectrometers produced (ARL and Baird Atomic); allow rapid multielement analysis of metals.
1948	First commercial mass spectrometers built by Consolidated Engineering Co. (USA) and Metropolitan Vickers Electric Co. (UK).
1949	Paul T. Gilbert creates a flame emission attachment for the popular Beckman DU Spectrophotometer.
1953	First quadrupole mass filter and quadrupole ion trap (Paul).
1955	Alan Walsh develops atomic absorption spectroscopy.
1956	Eugen Badarau described the first application of an induction plasma as a spectral source.
1956	First commercially available vacuum optical emission spectrometer (ARL Quantovac).
1958	Arthur L. Schalow and Charles H. Townes publish the article 'infrared and optical masers', describing the principles of the laser.
1959	M. Margoshes and B.F. Scribner describe the 'plasma jet'; for the first time the solution to be analysed was not in contact with the electrodes.
1960	Theodore Maiman (Hughes Research Laboratories) produces the first operational (ruby) laser.
1961	T.B. Reed uses a high power induction torch to grow refractory crystals. In this work, for the first time, he protected the torch wall with a stream of argon.
1961	Boris L'Vov develops the graphite furnace for atomic absorption spectroscopy.
1963	First computer-controlled optical emission spectrometers.
1963	Jan van Calker and Wilhelm Tappe report experiments with RF generators; reporting tests with a microwave induced plasma, capacitive coupled microwave plasma and inductively coupled plasma.
1963	R. Mavrodineanu and R.C. Hughes report tests on all possible plasma sources including a low power one at 30 MHz.
1963–1964	Stanley Greenfield, I.L. Jones and C.T. Berry publish on the annular inductively coupled plasma.
1964	H. Dunken, G. Pforr and W. Mikkeleit reported on the use of an ultrasonic nebulizer for liquid sample introduction into a plasma.
1965	R.H. Wendt and Velmer A. Fassel publish the first paper on plasma spectral excitation.

(Continued)

Table 5.1 (Continued)

Date	Event
1966	Max Amos and John Willis introduce the nitrous oxide–acetylene flame.
1967	W. Grimm invents the glow discharge.
1968	L. DeGalan, R. Smith and J. Winefordner publish theoretical studies in the ICP.
1969	R.S. Babington introduces the high solids nebulizer.
1972	First commercially available portable arc/spark optical emission spectroscopy introduced by Hilger Ltd.
1974	First commercially available ICP spectrometers introduced.
1974	Pneumatic nebulization used for the first time into an ICP (Scott, Fassel, Kniseley and Nixon).
1974	Pneumatic cross-flow nebulizer (Kniseley, Amenson, Butler and Fassel).
1974	Meinhard concentric nebulizer introduced.
1974	Alan L. Gray couples a Direct Current Plasma with a mass spectrometer.
1975	J. Robin and C. Trassy first use end-on observation of the ICP.
1978	Pneumatic concentric nebulizer (Scott).
1978	Yost and Enke publish paper on triple-quadrupole mass analyser.
1978	V-groove high solids nebulizer (Suddendorf, and Boyer).
1979	Adjustable cross-flow nebulizer.
1980	First paper on the coupling of an ICP with a mass spectrometer published (Houk, Fassel, Flesch, Svec and Gray).
1982	Glass Babington V-groove nebulizer (GMK).
1983	Sapphire V-groove nebulizer (Jarrell Ash).
1983–1984	Meinhard C- and K-type concentric nebulizer introduced.
1984	First commercial ICP–MS instruments launched (Sciex ELAN250; VG PlasmaQuad).
1984	First commercial ICP–MS sold to the UK's Ministry of Agriculture, Fisheries and Food, in Norwich (VG PlasmaQuad).
1984–1985	Burgener–Legere V-groove Nebulizer: first commercial Teflon nebulizer.
1986	Microconcentric nebulizer for direct injection into plasma introduced (Fassel, Rice and Lawrence).
1986	First commercial laser ablation system for ICP–MS (VG Laser Lab).
1987	Ceramic V-groove nebulizer (Glass Expansion, Veespray).
1988	VG Elemental launch the PQ2 ICP–MS instrument.
1988	CETAC ultrasonic nebulizers with desolvator.
1989	Perkin Elmer/Sciex launch the ELAN 500 ICP–MS instrument.
1989	First commercial magnetic sector high resolution ICP–MS (VG PlasmaTrace).

Date	Event
1989	Commercial magnetic sector high resolution ICP–MS (JEOL, Plasmax).
1980's	Cyclonic spray chambers.
1992	First commercially available spectrometer with a charge injection device (CID) solid state detector (IRIS from Therma Jarrell Ash).
1992	First commercial multiple collector ICP–MS, MC–ICP–MS (VG, Plasma 54).
1993	Meinhard high efficiency nebulizer.
1994	First benchtop ICP–MS (HP4500; Yokogawa Analytical/Hewlett-Packard).
1994	First commercial HR (high resolution) ICP–MS not based on an existing organic MS system (Finnigan MAT, Element).
1994–1995	Burgener 'parallel path' nebulizers.
1997	MC–ICP–MS with collision cell (Micromass, iso-Plasma-trace, subsequently re-branded, Isoprobe).
1997	CETAC microconcentric nebulizer.
1997	First quadrupole MS with a hexapole collision cell (Micromass, platform).
1997	MC–ICP–MS (Nu Instruments, Nu Plasma).
1997	Meinhard direct injection high efficiency nebulizer, DIHEN.
1998	HR–ICP–MS (Thermo Optek, Axiom).
1998	HR–ICP–MS (ThermoQuest, Element 2).
1999	First dynamic reaction cell ICP–MS (ELAN 6100 DRC).
1999	MC–ICP–MS (Thermo Fisher Scientific, Neptune).
1999	Total consumption microconcentric nebulizer for hyphenated chromatography – ICP (CETAC).
2000	Separation of $^{40}Ca^+$ and $^{40}Ar^+$ using a Fourier transform ion cyclotron resonance mass spectrometer (ICP–FT–ICR–MS, Barshik et al.).
2001	Burgener mira mist.
2002	MC–ICP–MS with larger geometry (Nu Instruments, Nu 1700).
2002	First GC–ICP–MS interface (Agilent).
2004	HR–ICP–MS (Nu Instruments, Attom).
2012	Triple-Quadrupole ICP–MS (Agilent, 8800 ICP–QQQ–MS).
2012	First paper on laser ablation – Mattauch–Herzog-ICP–MS.

the kinetic energy of the electrons and the *gas temperature* the kinetic energy of the atoms. The plasma temperature is also inhomogeneous, in that temperature variation occurs both radially and axially. This means that the plasma is not in local thermodynamic equilibrium and are indeed complex sources.

Table 5.2 Current commercial ICP–AES and ICP–MS instruments.[a]

Company	Instruments
Agilent	7800 or 7900 ICP–MS
	8900 ICP–QQQ–MS
	5110 ICP–AES
Analytik Jena	PlasmaQuant PQ9000/PQ9000 Elite ICP–AES
	PlasmaQuant MS/Elite ICP–MS
Horiba Scientific	Ultima Expert/Expert LT ICP–AES
Nu Instruments	Sapphire MC–ICP–MS
	Plasma 3 MC–ICP–MS
	Attom ES MC–ICP–MS
	Plasma 1700 HR–MC–ICP–MS
Perkin Elmer	NexION® 2000 or 1000 ICP–MS
	Avio® 500 or 200 ICP–AES
Shimadzu	ICPS 7500/8100/9800 ICP–AES
	2030 ICP–MS
Spectro	Arcos, Blue or Genesis ICP–AES
	Spectro ICP–SF–MS
Thermo Fisher Scientific	Single quadrupole (SQ–ICP–MS)
	Triple quadrupole (TQ–ICP–MS)
	HR–ICP–MS
	MC–ICP–MS
	iCAP 7200, 7400 or 7600 ICP–AES
TOFWERK	icpTOF (ICP–ToF–MS)

a) As of 1 August 2018.

The ICP is formed within the confines of three concentric fused quartz glass tubes of a plasma torch (Figure 5.1). Each concentric fused quartz tube has an entry point, with the intermediate (plasma) and external (coolant) tubes having tangentially arranged entry points and the inner tube consisting of a capillary tube through which the aerosol is introduced from the nebulization/spray chamber. Located around the outer fused quartz tube is the load coil. Conventionally, this was made of coiled copper tubing through which water was recirculated to aid cooling; an alternate format of two flat aluminium plates, without cooling water, acting as the load coil is also available.

Power input to the ICP is achieved through this copper (load or induction) coil, typically in the range 0.5–1.5 kW at a frequency of 27 or 40 MHz via an RF generator. The input power causes the induction of an oscillating magnetic field whose lines of force are axially orientated inside the plasma torch and

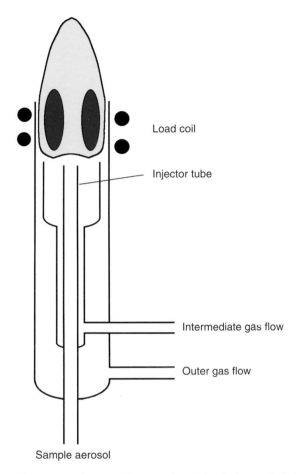

Load coil

Injector tube

Intermediate gas flow

Outer gas flow

Sample aerosol

Figure 5.1 Schematic diagram of an inductively coupled plasma torch.

follow elliptical paths outside the induction coil (Figure 5.2). At this point in time, no plasma exists. To initiate the plasma, the carrier gas flow is first switched off and a spark added momentarily from a Tesla coil, which is attached to the outside of the plasma torch by means of a piece of copper wire. Instantaneously, the spark, a source of 'seed' electrons, causes ionization of the argon carrier gas. This process is self-sustaining so that argon, argon ions and electrons co-exist within the confines of the plasma torch but protruding from the top in the shape of a bright white luminous 'bullet'.

The escaping high-velocity argon gas, causing air entrainment back towards the plasma torch itself, forms the characteristic 'bullet shape' of the ICP. To introduce the sample aerosol into the confines of the hot plasma gas (7000–10 000 K), the carrier gas is switched on: this punches a hole in the centre of the

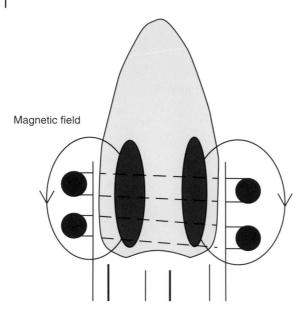

Magnetic field

Figure 5.2 Schematic representation of the formation of an inductively coupled plasma.

plasma, thus creating the characteristic 'doughnut' or toroidal shape of the ICP. In the conventional ICP system for atomic emission spectrometry, the emitted radiation can either be viewed laterally (side-on) or axially (end-on). Therefore, the element radiation of interest is 'viewed' either through the luminous plasma (laterally) or via the central channel (axially). The use of axial viewing is aimed at improving the efficiency of observation of the most useful zone in the ICP, that is, the central injector channel, while avoiding the surrounding plasma, which produces an intense and unwanted background. It is possible to calculate the optical solid angle of acceptance of the emission from the plasma using Eq. (5.1), whereby the solid angle of acceptance is calculated to be 0.03 sr [6]:

$$\Omega = \pi a^2 / 4 F^2 \tag{5.1}$$

where Ω = solid angle of acceptance, in steradians; a = effective aperture (35 mm) and F = focal length of lens (180 mm). The observed emission intensity also has a major contribution from the region within the depth of focus, calculated to be 20 μm at 400 nm using Eq. (5.2):

$$\delta = \pm 2\lambda \left(F / a \right)^2 \tag{5.2}$$

where δ = depth of focus and λ = wavelength. In effect, smoothing the spatial variations of intensity by summing the emission for all regions of the plasma within the angle of acceptance. Therefore, as the axially viewed plasma has a longer path length (i.e. depth of focus) this will result in superior detection limits (5–10-fold improvement), compared to side-on viewing (see Table 6.4). However, a compromise is required as in the axial viewed plasma there are increased spectral interference problems and matrix-induced interferences. Also, as the spectral observations are made through the much cooler tail plume of the plasma, self-absorption effects are common, resulting in reduced linear dynamic ranges [15]. However, it was necessary to design an interface that would permit the removal of the hot air/argon interaction zone and the efficient observation of the central channel. By including a cone to protect the collimating optical lens system from degradation a shear gas [16] was included between the tip of the plasma and the cone (Figure 5.3). The inclusion of a shear gas (Figure 5.3) meant that the matrix-induced interferences could be reduced [16].

Figure 5.4 compares the typical background emission characteristics observed for a conventional, laterally viewed plasma with that of an axially viewed plasma. The two major features of the spectra are the presence of many emission lines and a background continuum. The emission lines are mainly due to the source gas, namely argon, but also the presence of atmospheric gases, for example nitrogen, and the breakdown components of water, for example OH. The laterally viewed plasma has a higher background continuum than the axially viewed plasma. In either case, the background continuum is due to radiative recombination of electrons and ions ($M^+ + e^- \rightarrow M + h\nu$) and the radiation loss of energy by accelerated electrons (Bremsstralung radiation).

Issues can arise with the use of a silica torch in that after prolonged use or, if analysing high salt samples, devitrification of the glass can occur. This is

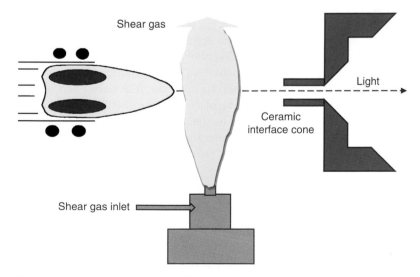

Figure 5.3 Axial (end-on) viewed ICP with shear gas.

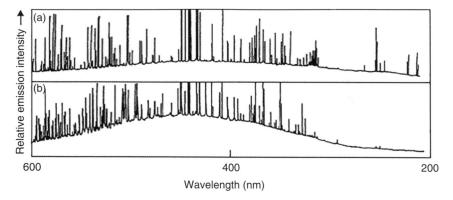

Figure 5.4 Comparison of the spectral features and background emission characteristics observed for (a) an axially viewed ICP and (b) a conventional, side-on viewed ICP [15]. *Source:* Reproduced with permission of the RSC.

observed by the silica glass becoming opaque, which will affect the analytical data for a laterally viewed ICP. It is therefore possible to use an alternate to the fused quartz outer tube and replace it with a ceramic, based on silicon nitride, tube. In addition, the silica quartz injector tube can also be replaced with either an alumina or sapphire injector tube. These alternate formats allow the ICP to be operated with difficult matrix solutions that contain hydrofluoric acid (HF), high dissolved solids and organic solvents.

5.4 Processes Within the ICP

During the introduction of an aqueous (or solid sample) into the plasma various fundamental processes take place that allow the ultimate visualization of the information via AES or MS. These processes are summarized in Figure 5.5. The mechanisms for excitation and ionization in the ICP have been widely discussed and it is believed that they take place because of collisions between analyte atoms with energetic electrons. The major benefit of the ICP source for AES and MS is its ability to vaporize, atomize, excite and ionize, efficiently and reproducibly.

5.5 Signal Processing and Instrument Control

The electronics used for signal processing in ICP–AES and ICP–MS are similar. In ICP–AES the electric current measured at the anode of the photomultiplier tube (PMT) or charged-coupled device is converted into a signal via

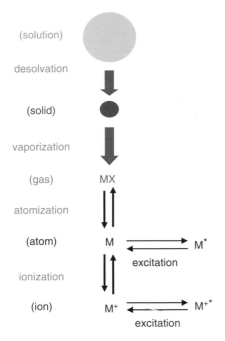

(solution)		
desolvation		
(solid)		
vaporization		
(gas)	MX	
atomization		
(atom)	M ⇄	M*
	excitation	
ionization		
(ion)	M⁺ ⇄	M⁺*
	excitation	

Figure 5.5 Processes that occur when a sample droplet is introduction into the plasma.

an analogue-to-digital converter. Similarly, in ICP–MS the electric current measured at the anode of the electron multiplier tube (EMT) is converted into a signal via an analogue-to-digital converter. This digital information is then processed, using sophisticated purpose designed software to provide useable spectral and intensity data.

Instrument computer control is an important aspect of ICP–AES and ICP–MS instruments. Use of sophisticated computer control allows the analyst to control the front-end (autosampler and ICP) through to the back-end (spectrometer that collects, manipulates and reports the analytical data). It is possible to have a fully automated instrument that allows unattended operation and self-protects in the case of an operating issue.

5.6   Summary

This chapter highlights the importance of plasma technology for elemental analysis. The fundamental aspects of the inductively coupled plasma are described.

References

1 Hittorf, W. (1884). *Wiedemanns Ann. D. Physik* 21: 90–139.
2 Greenfield, S., Jones, I.L.I., and Berry, C.T. (1964). *Analyst* 89: 713–720.
3 Wendt, R.H. and Fassel, V.A. (1965). *Anal. Chem.* 37: 920–922.
4 Gray, A.L. (1974). *Proc. Soc. Anal. Chem.* 11: 182–183.
5 Houk, R.S., Fassel, V.A., Flesch, G.D. et al. (1980). *Anal. Chem.* 52: 2283–2289.
6 Thomsen, V. (2006). *Spectroscopy* 21 (10): 32–42.
7 Gray, A.L. (1986). *J. Anal. At. Spectrom.* 1: 403–405.
8 Greenfield, S. (2000). *J. Chem. Ed.* 77 (5): 584–591.
9 Potter, D. (2008). *J. Anal. At. Spectrom.* 23: 690–693.
10 Griffiths, J. (2008). *Anal. Chem.* 80: 5678–5683.
11 Douthitt, C.B. (2008). *J. Anal. At. Spectrom.* 23: 685–689.
12 Munzenberg, G. (2013). *Int. J. Mass Spectrom.* 349–350: 9–18.
13 Ohls, K. and Bogdain, B. (2016). *J. Anal. At. Spectrom.* 31: 22–31.
14 Burgener Research Inc. (http://www.burgenerresearch.com).
15 Davies, R., Dean, J.R., and Snook, R.D. (1985). *Analyst* 110: 535–540.
16 Demers, D.R. (1979). *Appl. Spectrosc.* 33: 584–592.

6

Inductively Coupled Plasma–Atomic Emission Spectrometry

LEARNING OBJECTIVES

- To be able to calculate the energy, wavelength or frequency of a spectral line in appropriate units.
- To be able to define the terms 'absorption' and 'emission' (of radiation).
- To be able to explain and understand the importance of identifying a resonance line (wavelength).
- To be able to calculate the relative populations of the ground and excited states for a spectral line.
- To be able to define spectral line width.
- To appreciate the main types of line broadening in atomic spectroscopy.
- To be able to calculate Doppler and natural line widths.
- To describe the two different approaches in which an ICP can be viewed.
- To be able to calculate the limit of detection and background equivalent concentration in relation to ICP analysis.
- To describe the merits of sequential and simultaneous multi-element detection.
- To be able to recognize the different optical arrangements of spectrometers used for atomic emission spectrometry.
- To be able to calculate diffraction grating parameters; for example, resolution.
- To understand the benefits of using an Echelle spectrometer.
- To be able to describe the operation of a photomultiplier tube.
- To understand the importance of charge-transfer devices in atomic emission spectrometry.
- To be able to identify interferences and their remedies in atomic emission spectrometry.

Practical Inductively Coupled Plasma Spectrometry, Second Edition. John R. Dean.
© 2019 John Wiley & Sons Ltd. Published 2019 by John Wiley & Sons Ltd.

6.1 Introduction

This chapter will investigate the role and function of inductively coupled plasma (ICP) in atomic emission spectrometry.

6.2 Fundamentals of Spectroscopy

The most common emission source for spectroscopic measurement is the ICP. The emission arises from specific energy changes within an atomic system. The regions of the electromagnetic spectrum can be identified in terms of a wavelength and a frequency (Figure 6.1). A relationship exists that allows wavelength (λ) and frequency (f) to be determined, provided that one of the terms is known. Wavelength is normally expressed in units of metres (m) and frequency in cycles per second (s^{-1}) or hertz (Hz). The relationship is as follows:

$$c = f \cdot \lambda \tag{6.1}$$

where c is the velocity of light and approximates to 3.00×10^8 ms^{-1}.

As well as frequency and wavelength, electromagnetic radiation can also be expressed in terms of 'packets' of energy (E) called *photons* (or *quanta*). The energy of a photon can be expressed in terms of frequency, as follows:

$$E = h \cdot f \tag{6.2}$$

where h is the Planck constant (6.626×10^{-34} J).

By substitution of Eq. (6.1) into Eq. (6.2) we can obtain an expression related directly to the wavelength:

$$E = (h \cdot c) / \lambda \tag{6.3}$$

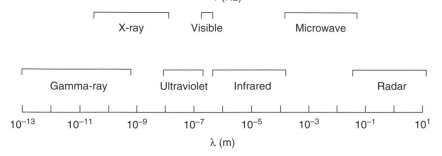

Figure 6.1 The electromagnetic spectrum.

6.2.1 Origins of Atomic Spectra

If an atom is supplied with enough (thermal) energy, the electron is raised from a low-energy level (e.g. ground state) to one with a higher energy (excited state). This is referred to as absorption. As the excited state is unstable, the electron returns to a lower-energy state (by inference, a more stable situation). This is referred to as emission. Both absorption and emission occur at certain selected wavelengths, frequencies, or energies (Figure 6.2).

At room temperature, all the atoms of a sample are in the ground state. For example, the single outer electron of sodium occupies the 3s orbital. [*Note*: the electron configuration for Na is $1s^2\ 2s^2\ 2p^6\ 3s^1$.] In a hot environment (i.e. the ICP), the sodium atoms are capable of absorbing radiation, such that electronic transitions from the 3s level to higher excited states can occur. These electronic transitions occur at specific wavelengths. Experimental observation of sodium identifies absorption peaks at 589.0, 589.6, 330.2 and 330.3 nm. By considering the energy-level diagram in Figure 6.3 for sodium, it is possible to identify that these wavelength doublets correspond to electronic transitions from the 3s level to either the 3p or 4p levels for 589.0/589.6 and 330.2/330.3 nm, respectively. While other electronic transitions are possible, the strongest, that is, the most intense, absorption spectrum occurs for electronic transitions from the ground state (3s) to upper levels. The wavelengths at which these transitions occur are called resonance lines.

In the hot environment of an ICP, the electron is easily excited to an upper-energy level. However, as the lifetime of the excited atom is brief (typically 10^{-8} seconds) its return to the ground state is accompanied by the emission of a photon of radiation. In Figure 6.4, the wavelength doublet at 590 nm (589.0 and 589.6 nm in Figure 6.3) represents the most intense emission lines for

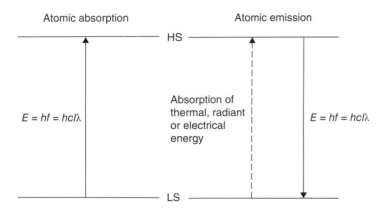

Figure 6.2 Schematic representation of atomic absorption and atomic emission energy transitions: LS, lower-energy state or ground state and HS, higher energy state.

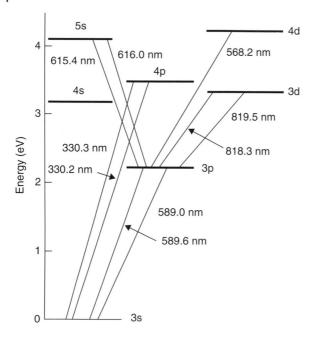

Figure 6.3 An energy-level diagram for sodium.

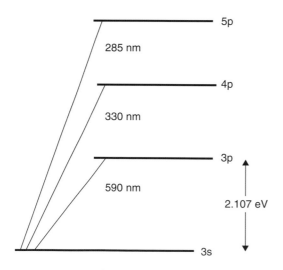

Figure 6.4 Simplified energy-level diagram for sodium.

sodium and is responsible for the yellow colour when sodium salts are introduced into a flame; for example, in a Bunsen burner. It is important to note that both the emission and absorption lines for sodium occur at identical wavelengths since the transitions involved are between the same energy levels.

Emission spectra can be further complicated by the presence of both band and continuous spectra. Band (or molecular) spectra arise from the excitation of molecular species in the hot environment of the source. Thus, it is common to observe band spectra from diatomic molecules. To differentiate these molecular species from the atomic species of interest, it is necessary to use a high-resolution spectrometer (see Section 6.4). Typical molecular species encountered are C_2 molecules (if organic solvents are introduced) and OH radicals. Their appearance can be troublesome, producing undesirable interference effects. Spectra continua that can arise from recombinations, such as electrons and ions that form atoms and Bremsstrahlung[1] in plasmas, generally cause elevation of the background against which the emission lines are measured.

6.2.2 Spectral Line Intensity

Spectral line intensities depend on the relative populations of the ground or lower electronic state and the upper excited state. The relative populations of the atoms in the ground or excited states can be expressed in terms of the Boltzmann distribution law, as follows:

$$N_1 / N_o = g_1 / g_o\, e^{(-\Delta E/kT)} \tag{6.4}$$

where N_1 is the number of atoms in the excited state, N_0 the number of atoms in the ground or lower state, g_1 and g_0 are the number of energy levels having the same energies for the upper (excited-) and lower (ground)-energy levels, respectively [*Note*: energy levels of the same energy are usually referred to as being degenerate.], ΔE the difference in energy between the lower-and upper-energy states, k the Boltzmann constant $(8.314\,\mathrm{J\,K^{-1}\,mol^{-1}})$ and T the temperature.

Again, using sodium as an example, there are two degenerate 3p energy levels (excited state) and a single ground state, 3s, producing a g_1 value of 2 and a g_0 value of 1. Simplification of this equation gives the following:

$$N_1 / N_o = 2e^{(-\Delta E/kT)} \tag{6.5}$$

It is then possible to calculate the ratio of the populated upper and ground transition states for the spectral transition at 589 nm at a typical plasma temperature of 7000 K. By using Eq. (6.3) it is possible to determine the energy of this spectral transition.

$$E = (6.626\times10^{-34}\ \mathrm{J.s}\times3.00\times10^8\ \mathrm{m.s^{-1}})\,/\,589\times10^{-9}\ \mathrm{m}$$

---

1 Continuous background emission – electromagnetic radiation arising from collision or deviation between fast-moving electrons or atoms (from the German: 'braking radiation').

$$E = 3.37 \times 10^{-19}\,\text{J}$$

or for 1 mol of photons 3.37×10^{-19} J $\times 6.022 \times 10^{23}$ mol^{-1} = 203 544 J mol^{-1}. By substitution into the revised Boltzmann equation:

$$N_1 / N_o = 2\exp(-203544\,\text{J}\,\text{mol}^{-1} / 8.314\,\text{J}\,\text{K}^{-1}\,\text{mol}^{-1} \times 7000\,\text{K})$$

$$N_1 / N_o = 2\,e(-3.50) \text{ or } 0.06$$

So, at a typical plasma temperature of 7000 K, 6% of the atoms are in the excited state.

6.2.3 Spectral Line Broadening

From the discussion so far, you might have the impression that emission line profiles are very narrow and occur at discrete wavelengths. This is true, but due to other processes that occur the observed spectral lines profiles are invariably broadened. Figure 6.5 shows the typical shape of a spectral line. The line width is defined as 'the width at half the peak height' ($\Delta\lambda_{1/2}$). Three main factors influence the line widths; namely natural, Doppler and pressure broadening.

The natural line width broadening is a consequence of the short life-time (approximately 10^{-8} seconds) of an atom in an excited state. In 1927, Werner Heisenberg postulated that 'nature places limits on the precision with which certain pairs of physical measurements can be made' [1]. Due to this 'Uncertainty Principle', the natural width of an emission line ($\Delta\lambda_N$) can be determined from the following expression:

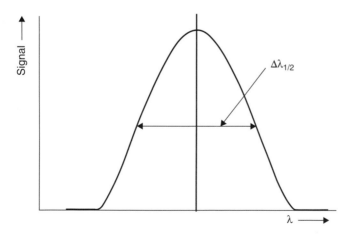

Figure 6.5 Typical shape of a spectral line.

$$\Delta\lambda_N = \lambda^2\, \Delta f\, /\, c \tag{6.6}$$

where λ is the wavelength of the emission line, Δf the uncertainty in the frequency of the emitted radiation [*Note*: Δf is equal to $1/\tau$, where τ is the lifetime of the excited state.] and c the speed of light.

On that basis, it is then possible to calculate the natural line width for a sodium emission line of wavelength 589 nm with an excited lifetime of 2.5×10^{-9} seconds. Using Eq. (6.6),

$$\Delta\lambda_N = ((589 \times 10^{-9}\, \text{m})^2\ (1/2.5 \times 10^{-9}\, \text{s})) / (3 \times 10^8\, \text{ms}^{-1})$$

$$\Delta\lambda_N = ((3.469 \times 10^{-13})\text{m}^2\ (4 \times 10^8\, \text{s}^{-1})) / 3 \times 10^8\, \text{ms}^{-1}$$

$$\Delta\lambda_N = 4.625 \times 10^{-13}\, \text{m}.$$

Therefore, the natural line width for sodium at 589 nm is 0.00046 nm or 0.46 pm.

It should be noted that the observed line width for sodium, at 589 nm, is some 10 times wider than the value found (0.46 pm). Therefore, other broadening processes must predominate.

The thermal motion of atoms in a gas (or plasma) introduces an additional broadening of the line profile; that is, Doppler broadening ($\Delta\lambda_D$). An equation that describes this broadening is given here, in units of metres:

$$\Delta\lambda_D = (2\lambda/c)\sqrt{(2RT/M)} \tag{6.7}$$

where λ is the wavelength of the emission, c the speed of light $(3 \times 10^8\, \text{ms}^{-1})$, R the gas constant $(8.314\, \text{J K}^{-1}\, \text{mol}^{-1})$, T the temperature and M the atomic mass of an atom.

The Doppler line width for the 589 nm spectral line of the sodium atom in a plasma of temperature 7000 K [*Note*: the atomic mass of sodium is 23 g mol^{-1}, but in SI units the value is 23×10^{-3} kg mol^{-1}.] can be determined using Eq. (6.7).

$$\Delta\lambda_D = ((2 \times 589 \times 10^{-9}\, \text{m}) / 3 \times 10^8\, \text{ms}^{-1}$$
$$\sqrt{((2 \times 8.314\, \text{J K}^{-1}\, \text{mol}^{-1} \times 7000\, \text{K}) / 23 \times 10^{-3}\, \text{kg mol}^{-1}))}$$

$$\Delta\lambda_D = (3.93 \times 10^{-15}\, \text{s})\sqrt{(116\ 396)\text{J}) / 23 \times 10^{-3}\, \text{kg}}$$

$$= (3.93 \times 10^{-15}\, \text{s})\sqrt{(5\ 060\ 696\, \text{J kg}^{-1})}$$

$$= 3.93 \times 10^{-15}\, \text{s} \times 2249.6\, \text{kg m}^2\, \text{s}^{-2}.\text{kg}^{-1})$$

$$= 3.93 \times 10^{-15}\, \text{s} \times 2249.6\, \text{m s}^{-1}$$

$$\Delta\lambda_D = 8.84 \times 10^{-12}\, \text{m or } 8.8\, \text{pm}$$

[*Note*: $J = \text{kg m}^2\, \text{s}^{-2}$.]

If you have carried out this calculation correctly, it should now be evident that the major contributor to the observed line width (0.009 nm) for sodium in a plasma is Doppler broadening.

Pressure broadening arises from collisions between the emitting species with other atoms or ions in the plasma. These collisions cause small changes in the ground-state energy levels and hence a subsequent small variation in the emitted wavelength. In plasmas, the collisions are between the atoms of interest and the argon of the plasma. [*Note*: this type of broadening is usually referred to as *Lorentz* broadening.] This results in significant broadening (similar, but slightly less than that obtained for Doppler broadening, i.e. typically 3 pm for sodium).

6.3 Plasma Spectroscopy

The concept of using an ICP as a source for atomic emission spectrometry is attributable to the independent work of Greenfield et al. [2] in the UK and Wendt and Fassel [3] in the USA. However, they were not the first to use an ICP torch – this was first reported by Reed [4] who used the technique for growing crystals under high-temperature conditions. While these early instruments would probably not be recognizable in the modern laboratory, the concept of inductively coupled plasma–atomic emission spectrometry (ICP–AES) as an elemental technique was born more than 55 years ago.

Light emitted from the plasma source is focused onto the entrance slit of a spectrometer by using a convex lens arrangement. However, two viewing modes are possible (see also Section 5.2). In 'side-on' or 'lateral' viewing, light from the plasma source is orthogonal to the central channel of the ICP, whereas in 'end-on' or 'axial' viewing light from the plasma source is coincident with the central channel of the ICP (Figure 6.6). Figure 5.4 compares the typical background spectra obtained by side-on and end-on viewing. An evaluation of radially and axially viewed ICPs has been reported [5]. Table 6.1 shows the experimental configurations for two plasma-based systems using axially and radially viewed configurations. Diagnostic tests were performed and these are summarized in Table 6.2. The limit of detection (LOD) and

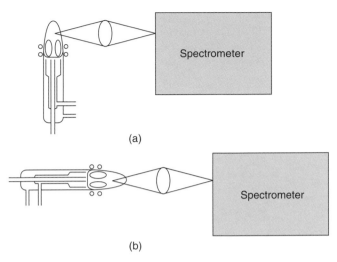

(a)

(b)

Figure 6.6 Schematic representations of the two viewing modes for ICPs: (a) axially (end-on) and (b) radially or lateral (side-on).

background equivalent concentration (BEC) were calculated by using the following definitions, respectively:

$$BEC = C_{RS} / SBR \qquad (6.8)$$

$$LOD = (3 \times BEC \times RSD) / 100 \qquad (6.9)$$

where C_{RS} is the concentration of a multi-element reference solution (20 mg l^{-1}), SBR the signal-to-background ratio ($= (I_{RS} - I_{blank})/I_{blank}$)), where I_{RS} and I_{blank} are the emission intensities for the multi-element and reference solutions, respectively, and RSD the relative standard deviation for 10 measurements of the blank (reference) solution.

The results obtained from the diagnostic tests are shown in Table 6.3. Robustness was quantified by measuring the Mg(II) and Mg(I) emission signals at a radiofrequency (RF) power (to the ICP torch) of 1.2 kW. The nebulizer gas flow rate was then adjusted until the maximum Mg(II)/Mg(I) ratio was achieved (0.90 and 0.70 l min^{-1} for the axially and radially viewed ICP systems, respectively). The resultant Mg ratio was then multiplied by 1.8 to correct for response intensities resulting from the use of an Echelle spectrometer with a charge-coupled device (CCD) detector [6]. It can be concluded that both axially and radially viewed plasma configurations offer similar figures of merit. Some improvements in sensitivity are noted (Tables 6.3 and 6.4) for the axially viewed plasma. However, the radially viewed plasma is ready for use in a shorter time

Table 6.1 Comparison of the experimental configurations used for axially and radially viewed ICP sources for AES [5]. Reproduced with permission of Elsevier.

Optical system	*Parameter*
Polychromator	Echelle grating plus CaF_2 cross-dispersing prism
Grating density groove	95 grooves mm^{-1}
Focal length	400 mm
Entrance slit	Height, 0.029 mm; width, 0.051 mm.
Detector	Peltier cooled CCD; 70 908 pixels spread across 70 non-linear arrays; wavelength range, 167–785 nm.
Sample introduction system	*Type*
Nebulizer	Concentric
Spray chamber	Cyclonic
Plasma conditions	*Parameter*
Frequency	40 MHz
RF power	1.2 kW
Plasma gas flow rate	15.0 l min^{-1}
Auxiliary gas flow rate	1.5 l min^{-1}
Nebulizer gas flow rate	0.8 l min^{-1}
Sample flow rate	0.8 ml min^{-1}
Torch injector tube diameter	2.3 mm (axially viewed ICP); 1.4 mm (radially viewed ICP)
Observation height[a]	13 mm
Emission lines (examples)[b]	Ar(ı) 404.442; Ar(ı) 404.597; Ba(ıı) 230.424; Ba(ıı) 455.403; Mg(ıı) 280.264; Mg(ı) 285.208; Ni(ıı) 231.604

a) Only for the case of a radially viewed ICP.
b) In nm.

('warm-up' time) than the axially viewed plasma. Despite these differences, the analytical performance of each system was not compromised. Both ICP systems were then applied to the analysis of two certified reference materials that had previously been microwave-digested in a closed-vessel system using nitric acid and hydrogen peroxide, to assess accuracy and precision. The results, shown in Table 6.5, indicate that both systems can provide accurate and precise results.

6.4 Spectrometers

The spectrometer is required to separate the emitted light into its component wavelengths. In practice, two options are available. The first of these involves

Table 6.2 Diagnostic procedures used for axially and radially viewed ICP sources for AES [5]. Reproduced with permission of Elsevier.

'Figure of merit'	Parameter[a),b)]
UV spectral resolution	Profile of Ba(II) 230 nm line
Visible spectral resolution	Profile of Ba(II) 455 nm line
Robustness	Mg(II) 280/Mg(I) 285 nm ratio
Short-term stability	RSD for Mg(I) 285 nm emission signal ($n = 15$)
Long-term stability	RSD for Mg(I) 285 nm emission signal ($n = 8$; $t = 2$ h)
Sensitivity	LOD for Ni(II) 231 nm line
'Warm-up' time	RSDs for Ar, Ba and Mg emission lines

a) RSD, relative standard deviation.
b) LOD, limit of detection.

Table 6.3 Results obtained from the diagnostic tests carried out on axially and radially viewed ICP sources for AES [5]. Reproduced with permission of Elsevier.

'Figure of merit'	Axially viewed[a)]	Radially viewed[a)]
UV spectral resolution	8 pm	8 pm
Visible spectral resolution	30 pm	30 pm
Robustness[b)]	10.6	13.7
Short-term stability	0.70% RSD	0.60% RSD
Long-term stability	1.5% RSD	1.4% RSD
Sensitivity[c)]	0.23 µg l^{-1}	3.8 µg l^{-1}
'Warm-up' time[d)]	20 min	10 min

a) RSD, relative standard deviation.
b) Multiplied by 1.8.
c) See also Table 6.4.
d) Emission intensity deviations less than 5%.

Table 6.4 Limit of detection (LOD) and background equivalent concentration (BEC) data for Ni(II) (231.604 nm) in 0.14 mol l^{-1} of HNO_3 and 1000 mg l^{-1} of Cr media for axially and radially viewed ICP sources for AES [5]. Reproduced with permission of Elsevier.

Media	BEC (µg l^{-1})		LOD (µg l^{-1})	
	Axially	Radially	Axially	Radially
HNO_3	11	301	0.23	3.8
Cr	4	741	0.31	7.6

Table 6.5 Analysis of two certified (standard) reference materials using axially and radially viewed ICP–AES [5]. Reproduced with permission of Elsevier.

Element (concentration)	Axially viewed	Radially viewed	Certificate value
NIST 1515 Apple Leaves[a]			
Calcium (wt%)	1.32 ± 0.06	1.46 ± 0.03	1.526 ± 0.015
Copper ($mg\,kg^{-1}$)	4.98 ± 0.23	6.14 ± 0.06	5.64 ± 0.24
Iron ($mg\,kg^{-1}$)	61.9 ± 1.4	66.5 ± 4.3	83 ± 5
Magnesium (wt%)	0.241 ± 0.010	0.247 ± 0.003	0.271 ± 0.008
Manganese ($mg\,kg^{-1}$)	44.5 ± 1.1	48.5 ± 2.5	54 ± 3
Zinc ($mg\,kg^{-1}$)	10.9 ± 0.3	19.6 ± 2.7	12.5 ± 0.3
NIST 1577b Bovine liver[a]			
Calcium ($mg\,kg^{-1}$)	146 ± 12	107 ± 5	116 ± 4
Copper ($mg\,kg^{-1}$)	142 ± 1	148 ± 1	160 ± 8
Iron ($mg\,kg^{-1}$)	164 ± 4	156 ± 4	184 ± 15
Magnesium ($mg\,kg^{-1}$)	546 ± 5	523 ± 4	601 ± 28
Manganese ($mg\,kg^{-1}$)	7.20 ± 0.58	9.14 ± 0.75	10.5 ± 0.17
Zinc ($mg\,kg^{-1}$)	114 ± 1	110 ± 3	127 ± 16

a) NIST, National Institute of Science and Technology.

a capability to measure one wavelength, corresponding to one element at a time, while the second allows multi-wavelength or multi-element detection. The former is known as sequential analysis or sequential multi-element analysis if the system is to be used to measure several wavelengths one at a time, while the latter is termed simultaneous multi-element analysis. The typical wavelength coverage of a spectrometer for atomic emission spectrometry (AES) is between 167 nm (Al) and 852 nm (Cs). [*Note*: Below 190 nm purged optics (using N_2) and a vacuum spectrometer are required due to the absorption of oxygen. However, the typical operation of most spectrometers is between 190 and 450 nm.]

Separation of the light into its component wavelengths is achieved in all modern instruments using a diffraction grating. The latter consists of a series of closely spaced lines ruled or etched onto the surface of a mirror. Most gratings for AES have a line, or groove, density between 600 and 3200 lines mm^{-1}. When light strikes the grating, it is diffracted at an angle that is dependent on the grating equation, as follows:

$$n\lambda = d\sin\phi \qquad (6.10)$$

where n is the spectral order, λ the wavelength, d the distance between a line or groove on the grating and ϕ is the angle of a groove.

By using Eq. (6.10), it is possible to calculate the expected wavelength of a spectral line. For example, the wavelength of a spectral emission line, in the first order, with a groove density of 1200 lines mm^{-1} and an angle, ϕ, of 20°. As the groove density is 1200 lines mm^{-1} this would equate to one groove every 0.00083 mm (or 830×10^{-9} m). By using Eq. (6.10) the wavelength would be:

$$n\lambda = 830 \times 10^{-9} \, m \times \sin 20$$

$$\lambda = 830 \times 10^{-9} \, m \times 0.3420$$

$$\lambda = 283 \times 10^{-9} \, m \, or \, 283 \, nm$$

that, for a second order emission line, would occur at 141.5 nm.

Interference or 'ghost' images, from overlapping wavelengths can occur. To prevent spectral overlap, it is possible to use blazed reflection gratings. In this situation, the grooves are ruled at a specified angle (known as the blaze angle) and appear as a 'saw-tooth' pattern. Thus, it is possible to have a blazed diffraction grating that is more efficient in a specific wavelength region. The typical arrangement of the diffraction grating is shown in Figure 6.7; the exception to this is the Echelle grating, which is described in Section 6.4.2. The resolution of the grating is related to the spectral order (m) and the total number of grooves (N), as follows:

$$R = mN \tag{6.11}$$

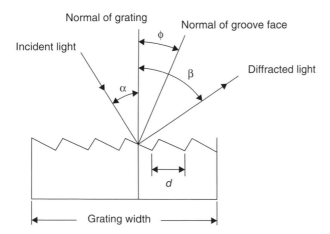

Figure 6.7 Schematic diagram of a blazed grating: d, distance between grooves; ϕ, angle of a groove (blaze angle); α, angle of incidence and β, angle of reflection.

while the resolving power, Δλ, is defined as the wavelength divided by the resolution.

Using Eq. (6.11), it is then possible to calculate (i) the resolution for a conventional grating ruled with 1200 lines mm^{-1} with a width of 52 mm in the first order, and (ii) the resolving power at 300 nm.

$$R = 1 \times (1200\,\text{mm}^{-1} \times 52\,\text{mm})$$

$$R = 62\,400$$

[*Note*: notice there are no units.]

With a resolving power, at a wavelength of 300 nm, of:

$$\Delta\lambda = 300\,\text{nm} / 62\,400$$

$$\Delta\lambda = 0.00481\,\text{nm}$$

6.4.1 Sequential Spectrometer

A sequential spectrometer is the lower-cost option for AES. This typically consists of entrance and exit optics, a diffraction grating and a single detector. A sequential spectrometer has the advantage of flexibility in terms of wavelength coverage. Selection of the desired wavelength is achieved by rotation of the grating within its spectrometer mounting. This rotation can be achieved manually, or more typically in modern instruments, by computer control.

The major disadvantage of operating with a sequential spectrometer is the inability to monitor two wavelengths at the same time. Ultimately, this will affect the precision of the data as any internal standard (an element that is not present in the sample) will not be monitored. The inclusion of an internal standard reduces any potential interferences from, for example, a change in viscosity from sample to sample. In practice, however, modern instruments have software control that will allow two wavelengths to be monitored, thus allowing 'pseudo-wavelength' information to be obtained.

The Czerny–Turner configuration is the most common spectral mounting for sequential AES [7]. The optical layout of a spectrometer that incorporates these features is shown in Figure 6.8.

6.4.2 Simultaneous Spectrometers

One of the major advantages of AES is the ability to perform simultaneous multi-element analysis. In simultaneous analysis, many wavelengths or

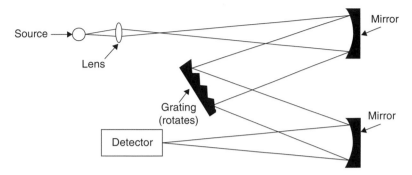

Figure 6.8 Schematic diagram of the optical layout of a spectrometer that incorporates the Czerny–Turner spectral mounting (configuration).

elements (typically, 20–70) can be monitored at the same time. More than one wavelength may be specified for each element if, for example, spectral interference from another element is known to occur. The limitation of such a system is that the exit slits are pre-set and this allows no flexibility if another element and/or wavelength is required to be analysed. This approach has traditionally been carried out by using a polychromator. The Paschen–Runge mounting (invented in the early twentieth century) is the most commonly used polychromator. The grating, entrance slit and multiple exits slits are fixed around the periphery of what is known as a 'Rowland circle' (invented in the late nineteenth century). The grating is concave in appearance and does not rotate. The optical layout of a spectrometer that incorporates these features is shown in Figure 6.9.

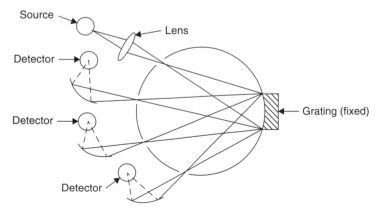

Figure 6.9 Schematic diagram of the optical layout of a spectrometer that incorporates the Paschen–Runge spectral mounting (configuration).

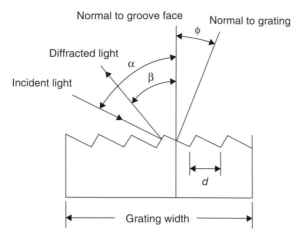

Figure 6.10 Schematic diagram of an Echelle grating: *d*, distance between grooves; ϕ, angle of a groove (blaze angle); α, angle of incidence and β, angle of reflection.

The major alternative approach for simultaneous multi-element analysis is the Echelle spectrometer [8]. While this spectrometer was originally only commercially available for the direct-current plasma (DCP), it has become increasingly important because of the special spectral features that are inherent in its design. The major component difference of the Echelle spectrometer is the grating. Unlike the previous design outlined, this grating utilizes the spectral order (recall the grating equation (6.10) for maximum wavelength coverage). A typical Echelle grating is ruled with only 50–100 lines or grooves per mm. The resolution of a diffraction grating is directly related to the groove density (n) and the spectral order (M) (Eq. (6.11)). In this situation, however, instead of using a grating with many grooves, the resolution is improved by increasing the blaze angle and spectral order. The arrangement of an Echelle grating is shown in Figure 6.10. If you compare this with the blazed grating shown in Figure 6.7, it is obvious that the light is reflected off the 'short-side' of the grating (and not the 'long-side', as in Figure 6.7). Therefore, the blaze angle is greater than 45°. The advantages of using this method to improve spectral resolution can be seen in Table 6.6.

However, to prevent overlapping of spectral orders a secondary dispersion is required. This is typically carried out by using a prism. If the latter is placed so that its separation occurs perpendicular to the diffraction grating, a two-dimensional spectral 'map' is produced (Figure 6.11). The spectral 'map' generated is thus sorted into spectral order vertically and wavelength horizontally (Figure 6.12). A typical optical layout of the Echelle spectrometer is shown in Figure 6.13.

Table 6.6 Comparison of the spectral features of a conventional diffraction grating and an Echelle grating.

Parameter	Conventional grating	Echelle grating
Focal length (m)	0.5	0.5
Groove or line density (lines mm^{-1})	1200	79
Diffraction angle	10° 22′	63° 26′
Width (mm)	52	128
Spectral order[a]	1	75
Resolution	62 400	758 400
Resolving power (nm)[a]	0.00481	0.000396

a) At 300 nm.

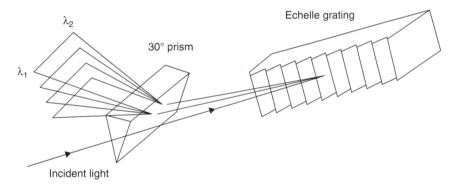

Figure 6.11 Schematic representation of the two-dimensional dispersion obtained by using a prism in conjunction with an Echelle grating.

6.5 Detectors

After wavelength separation has been achieved, it is necessary to 'view' the spectral information. Two different types of detectors are used in AES. Up until the 1990s, the most common detector was the photomultiplier tube (PMT); however, more recently this has largely been replaced by multi-channel detectors based on charge-transfer device (CTD) technology. Both a CCD and a charge-injection device (CID) are used for multi-channel detection in AES.

For the different spectrometer configurations described in the previous section, either type of detector can be used. In each case, the detector (PMT or CTD) is mounted behind the exit slit of the spectrometer. Thus, for the

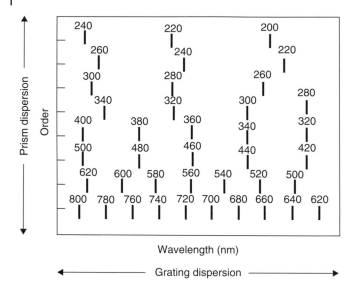

Figure 6.12 Spectral map generated by the Echelle spectrometer, using the arrangement shown in Figure 6.11.

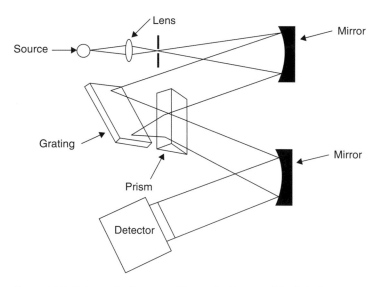

Figure 6.13 Schematic diagram of the optical layout of the Echelle spectrometer.

Czerny–Turner mounting (Figure 6.8) a single detector is required, while for the Paschen–Runge mounting (Figure 6.9) with 30 exit slits, 30 detectors are required. The latter case obviously adds to the cost and complexity of a polychromator system. However, the major advantage of CTD technology, that is, multi-wavelength detection with a single detector, has most effectively been exploited by using an Echelle spectrometer. The capability of an Echelle spectrometer to generate a two-dimensional spectral 'map', coupled with a sensitive multi-wavelength detector (i.e. a CCD or CID), provides a complete fingerprint of a sample.

6.5.1 Photomultiplier Tube

The PMT is a device that converts incident light into a current [9]. This is achieved by a series of steps that can be outlined by consideration of Figure 6.14. Incident light passes through the silica window and strikes the photocathode. The latter consists of an easily ionized material such as an alloy of two (or three) alkali metals with antimony. The exact composition of the photocathode affects the wavelength coverage of any detector. For example, in Figure 6.15 the use of the bialkali Sb–K–Cs provides a greater wavelength coverage than does the use of the high-temperature bialkali Na–K–Sb. In this manner, it is possible to select a PMT that has optimum response characteristics for a wavelength range. This feature can easily be exploited in a polychromator where an individual PMT is used to monitor only one wavelength.

Often the PMT that gives the widest spectral wavelength coverage (190–800 nm), that is trialkali (Na-K-Sb-Cs), is used in a monochromator.

A suitably energetic incident photon of light causes an electron to be emitted from the cathode surface. This process is known as the photoelectric effect,

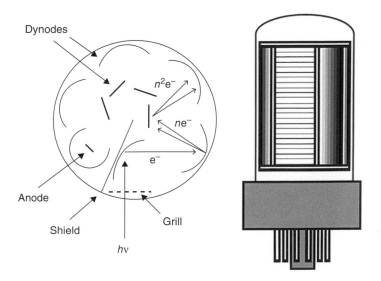

Figure 6.14 Schematic representation of the operation of a photomultiplier tube [10]. Reproduced with permission of John Wiley & Sons.

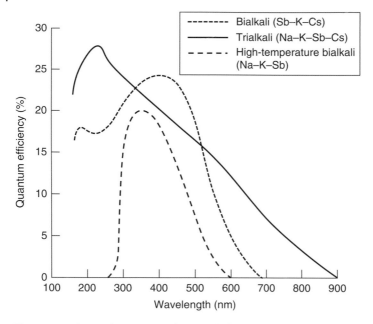

Figure 6.15 Spectral responses of selected photocathode materials.

while the yield per incident photon is called the quantum efficiency. A typical value for the latter is between 10 and 25% at 650–340 nm. By a series of focusing electrodes, the emitted electron is directed towards the dynode chain, where the latter acts to multiply this single electron into many electrons. This is achieved because a single electron striking dynode 1 will emit at least two secondary electrons. These electrons will then strike dynode 2 and for each electron that strikes the second dynode, at least a further two electrons are emitted and so on. In this manner, a cascade of electrons is produced. The exact number of electrons depends on the length of the dynode chain, which typically consists of 9–16 dynodes. This amplification of electrons by the dynode chain is known as the 'gain'. A typical gain for a nine-dynode PMT is 10^6. The final step is to collect the electrons at the anode. The electrical current that is measured at the anode is proportional to the amount of light that struck the photocathode. This current is then converted into a voltage signal that is then transformed via an analogue-to-digital (A/D) converter to a suitable computer for processing purposes.

6.5.2 Charge-Transfer Devices

The utilization of so-called CTDs is perhaps the most significant advance in detector technology for AES. CTDs offer a high sensitivity and a wide

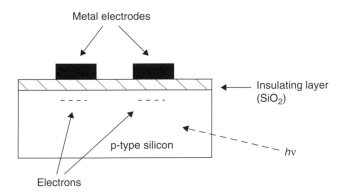

Figure 6.16 Schematic representation of the operation of a charged-coupled device [10]. Reproduced with permission of John Wiley & Sons.

wavelength coverage; that is, UV to visible spectra. Two common forms are available, the CCD and the CID. The first CCD was invented in 1969 [11] while the CID in 1984 [12]. Their main application has been as detectors for the Echelle spectrometer, although they have been used on polychromator systems using linear CCD arrays around the Rowland circle (see previously). The compact nature of the spectral 'map' generated by the Echelle spectrometer can be focused onto a single CTD. All CTDs are semiconductor devices consisting of a series of cells or pixels that accumulate charge when exposed to light. The amount of stored (accumulated) charge is then a measure of the amount of light to which a pixel has been exposed. A CTD is an array of closely spaced metal–insulator–semiconductor diodes formed on a wafer of semiconductor material. In operation, a CTD must be exposed to light for a specific period, and then 'read'. During this 'read' time, the detector is not exposed to the incoming light. Two common forms are available; that is, the CCD and the CID. In the case of a CCD (Figure 6.16), light from the plasma source is gathered for a specified time and then 'read out' on a 'row-by-row' basis, so allowing the accumulated charge to be transferred from pixel to pixel until it reaches the 'read-out' amplifier. A CID is slightly different in its operation in that it has additional circuitry that allows for the 'read-out' of individual pixels; that is, the CID pixels can be interrogated individually at any time during exposure to the plasma source.

The major advantages of CTDs are as follows:

- Flexibility in analytical line (wavelength) selection.
- Use of several lines (wavelengths) for the same element to extend the linear dynamic range.
- Use of several lines (wavelengths) for the same element to improve accuracy and to identify potential matrix or spectral interferences.

- Potential for qualitative analysis.
- Plasma-based diagnostic studies; for example, temperature measurement (see Section 4.1).

However, CTDs also suffer from some limitations relating to their function and operation [13].

6.6 Interferences

Spectral interferences for AES can be classified into two main categories; that is, spectral overlap and matrix effects. *Spectral interferences* are probably the most well known and best understood. The usual remedy to alleviate a spectral interference is to either increase the resolution of the spectrometer or to select an alternative spectral emission line. Three types of spectral overlap can be identified, as follows (Figure 6.17):

1) direct wavelength coincidence from an interfering emission line;
2) partial overlapping of the emission line of interest from an interfering line in close proximity;
3) the presence of an elevated or depressed background continuum.

Type (1) and (2) interferences can occur because of an interfering emission line from another element, the argon source gas or impurities within or entrained in the source, for example molecular species, such as OH and N_2. Extensive work has meant that wavelength coincidence (type 1) is well characterized for the ICP, for example, Cd (at 228.802 nm) and As (at 228.812 nm); Zn (at 213.856 nm) and Ni (at 213.858 nm). Elimination of the type (2) interference is usually only possible by an improvement in resolution. As this may not be possible on a routine basis, mathematical models can be used to try and correct for this type of interference. The only certain remedy, however, is to select an interference-free wavelength for the selected element. The type (3) interference can be corrected for by measurement of the background on either side of the wavelength of interest. Provided that no significant fine structure is present on the background, this method of correction should prove to be satisfactory.

Matrix interferences are often associated with the sample introduction process. For example, pneumatic nebulization can be affected by the dissolved-solids content of the aqueous sample that affects the uptake rate of the nebulizer and, hence, the sensitivity of the determination. Matrix effects that are encountered in the plasma source have also been documented. Typically, this involved the presence of easily ionizable elements (EIEs), for example alkali metals, within the plasma source. Some specific work has investigated the effect of EIEs on signal suppression or enhancement for the ICP source. The effects are greatest in the DCP source where the addition of lithium or barium salts is used as a buffer to reduce the problem of signal enhancement.

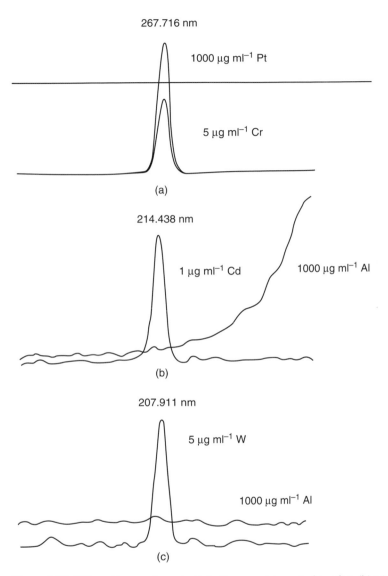

Figure 6.17 Different types of spectral interferences: (a) spectral overlap; (b) wing overlap and (c) background shift.

6.7 Summary

The use of the inductively coupled plasma for atomic emission spectrometry was highlighted in this chapter, and the fundamentals of spectroscopy relating to its use with an ICP were discussed. Of current interest are the optical viewing positions of the plasma relative to the spectrometer. The main spectrometer designs for ICP–AES were described, with particular emphasis placed on simultaneous and sequential systems. The use of charge-transfer technology for optical detection in AES was also emphasized.

References

1 Heisenberg, W. (1927). *Z. Phys. (in German)* 43 (3–4): 172–198.
2 Greenfield, S., Jones, L.I., and Berry, C.T. (1964). *Analyst* 89: 713–720.
3 Wendt, R.H. and Fassel, V.A. (1965). *Anal. Chem.* 37: 920–922.
4 Reed, T.B. (1961). *Appl. Phys.* 32: 821–824.
5 Silva, F.V., Trevizan, L.C., Silva, C.S. et al. (2002). *Spectrochim. Acta* 57: 1905–1913.
6 Dennaud, J., Howes, A., Poussel, E., and Mermet, J.M. (2001). *Spectrochim. Acta* 56: 101–112.
7 Czerny, M. and Turner, A.F. (1930). *Z. Phys.* 61: 792–797.
8 Nagaoka, H. and Mishima, T. (1923). *Astrophys. J.* 57: 92–97.
9 Iams, H. and Salzberg, B. (1935). *Proc. Inst. Radio Eng.* 23: 55–64.
10 Hou, X. and Jones, B.T. (2006). Inductively coupled plasma/optical emission spectrometry. In: *Encyclopedia of Analytical Chemistry* (ed. R.A. Meyers), 9468–9485. Chichester: Wiley.
11 Boyle, W. and Smith, G. Three-dimensional charge coupled devices, US Patent 3796927 A (16 December 1970).
12 Kelly, A.J., Charge injection device, US Patent 4630169 A (4 September 1984).
13 Mermet, J.M. (2005). *J. Anal. At. Spectrom.* 20: 11–16.

7

Inductively Coupled Plasma–Mass Spectrometry

LEARNING OBJECTIVES

- To understand the relationship between atomic number, atomic weight and isotopes.
- To appreciate the variation in the first and second ionization energies for selected elements.
- To be able to describe the operation of an inductively coupled plasma–mass spectrometer system.
- To understand the requirements of the interface for ICP–MS.
- To understand the principle of operation of a quadrupole mass spectrometer.
- To appreciate the different methods of signal monitoring in mass spectrometry.
- To appreciate the significance of using a sector-field mass spectrometer.
- To be aware of other types of mass spectrometers for ICP–MS.
- To be able to describe the operation of an electron multiplier tube.
- To be able to identify isobaric, molecular and matrix interferences in ICP–MS.
- To appreciate the different approaches for alleviating molecular interferences.
- To be aware of the different reactions that are possible in a collision/reaction cell for removal of spectroscopic interferences in ICP–MS.
- To be able to perform an isotope dilution analysis calculation.
- To be able to interpret ICP–MS spectra.

7.1 Introduction

The development and applications related to an inductively coupled plasma (ICP) coupled to a mass spectrometry are considered. Significant instrumental developments have taken place in ICP–MS that allow interferences to be overcome or avoided.

Practical Inductively Coupled Plasma Spectrometry, Second Edition. John R. Dean.
© 2019 John Wiley & Sons Ltd. Published 2019 by John Wiley & Sons Ltd.

7.2   Fundamentals of Mass Spectrometry

Mass spectrometry is a technique for measuring the molecular weight of elements or compounds. Most applications of mass spectrometry are for organic compounds, however, in this present context we consider its use for elemental analysis.

7.2.1   Some Terminology

Each element in the Periodic Table is composed of atoms, which themselves are composed of a nucleus (a mixture of protons and neutrons) and electrons. So, while a proton is positively charged and an electron negatively charged, the atom itself is neutral; that is, it has no charge. On that basis, the atom must consist of equal numbers of protons and electrons. It is possible for the same element, however, to have a different mass. In this situation, we refer to the *isotopes* of an element. So, for an element with isotopes, it must have a different number of neutrons.

For example, chlorine (Cl) is made up of 17 protons and 17 electrons. However, it can exist as two isotopes, each with a different mass. The lighter isotope contains 18 neutrons and is represented as $^{35}_{17}$ Cl, while the heavier isotope is represented as $^{37}_{17}$ Cl and contains 20 neutrons. The subscript represents the atomic number of chlorine, that is, the number of protons, while the superscript represents the atomic weight, that is, the total number of protons and neutrons. Therefore, the two isotopes of chlorine have atomic masses of 35 (actual 34.968852) and 37 (actual 36.965303) amu (atomic mass unit). But what about their abundances? If the lighter isotope of chlorine has an abundance of 75.77% while that of the heavier isotope is 24.23%, what is the resultant overall atomic weight of chlorine?

$$\text{Atomic weight of chlorine} = (34.97 \times 75.77/100) + (36.97 \times 24.23/100)$$
$$= (26.50) + (8.95).$$

Therefore, the atomic weight of chlorine is 35.45.

Ionization is the process whereby an electron can be removed from a neutral atom by applying an external source of energy, for example, the ICP. The resultant ion (a single positive cation if one electron is removed) has the same atomic mass as the original isotope of the element, as the mass of the electron is negligible (9.110×10^{-31} g when compared to the mass of a proton, 1.673×10^{-27} g, or that of a neutron, 1.675×10^{-27} g). The energy required to remove an electron is called the ionization energy. By application of further energy, a second electron can be removed from the resultant ion. Examples of selected element ionization energies are presented in Table 7.1 where all first ionization energies

Table 7.1 Selected ionization energies (in eV) for a range of elements.

Element	Symbol	Atomic number	First ionization energy	Second ionization energy
Lithium	Li	3	5.392	75.622
Sodium	Na	11	5.139	47.292
Potassium	K	19	4.341	31.811
Rubidium	Rb	37	4.177	27.499
Zinc	Zn	30	9.394	17.960
Cadmium	Cd	48	8.993	16.904
Mercury	Hg	80	10.437	18.752
Silicon	Si	14	8.151	16.339
Germanium	Ge	32	7.899	15.93
Tin	Sn	50	7.344	14.629
Lead	Pb	82	7.416	15.04

are in the approximate range 4–10 eV and all second ionization energies are higher (>14 eV).

All mass spectrometers can separate ions based on their mass/charge ratio, m/z; that is, the atomic mass of the element divided by its charge. It is normal that the charge on the element is a single positive one, that is, a cation; therefore, $z = 1$. [*Note*: the symbol for charge is sometimes represented as 'e' rather than 'z'.]

A range of mass spectrometers can be used for inductively coupled plasma–mass spectrometry (ICP–MS) (see Section 7.4). However, an important characteristic of any mass spectrometer is its ability to separate ions that have similar m/z ratios; that is, resolution (R). This is defined by considering two adjacent m/z ratios, that is, an isotope with a mass of 'm' and a second isotope (can be the same element or a different one) with a mass of '$m + \Delta m$' (Figure 7.1).

$$R = m / \Delta m \qquad (7.1)$$

7.3 Inorganic Mass Spectrometry

The coupling of a plasma source (capillary arc) to a mass spectrometer was first reported in 1974 [1]. However, Gray and co-workers soon noticed that the use of a capillary arc was a limited ion source. The obvious replacement ion source was the ICP, which at that time had found acceptance in research laboratories as a source for atomic emission spectroscopy. The coupling of an ICP with a

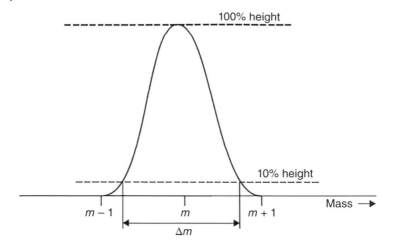

Figure 7.1 Resolution, as defined in mass spectrometry.

mass spectrometer was first reported in 1980, involving collaboration between Gray and the Ames Laboratory, USA [2]. For historical accounts of the development of ICP–MS, the reader is directed towards the work of Gray [3, 4]. Now, some 40 years later, the application of ICP–MS is routine.

In contrast to other sources for inorganic mass spectrometry, the ICP offers several advantages, not least the ability to analyse samples rapidly. This major advantage coupled with the high degree of sensitivity offered by ICP–MS are the unique features that have allowed this technique to evolve into the major analytical technique for elemental analysis post-2000. The major instrumental development required to establish this technique was the efficient coupling of an ICP, operating at atmospheric pressure, with a mass spectrometer, which operates under high vacuum (Figure 7.2). The development of a suitable interface held the key to the establishment of the technique.

An important feature of a mass spectrometer is its ability to measure isotope ratios. The importance of this feature is readily observed when you consider that approximately 70% of the elements in the Periodic Table have stable (non-radioactive) isotopes. The ability to measure isotope ratios has two major benefits, as follows:

- The use of stable isotopes for *tracer* studies, for example, monitoring the absorption of nutrients in people without the need for radio-labelled isotopes.
- The use of enriched stable isotopes for quantitative analysis; that is, isotope dilution analysis (IDA).

Figure 7.3 shows the mass spectra of lead isotopes, where Figure 7.3a displays the mass spectrum of (stable) lead, together with its relative abundances, while

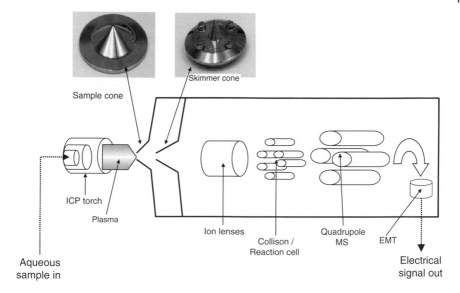

Figure 7.2 A schematic diagram of an inductively coupled plasma–mass spectrometer (ICP–MS) system.

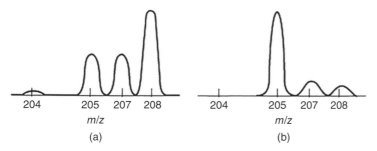

Figure 7.3 Mass spectra of lead isotopes, showing their relative abundances: (a) stable lead and (b) enriched lead.

Figure 7.3b shows a typical mass spectrum obtained for enriched lead, along with its relative abundances. Further details of IDA can be found in Section 7.7.

7.3.1 The Ion Source: ICP

The ion source is the ICP. Its operation and formation have been described earlier (see Section 5.3). In atomic emission spectroscopy, the ICP can be viewed either axially or radially (see Figure 6.6), while in mass spectrometry the ICP torch is positioned horizontally so that the ions can be extracted from the ICP directly into the mass spectrometer (and as the mass spectrometer is

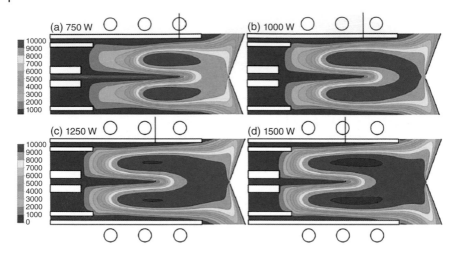

Figure 7.4 Modelling the inductively coupled plasma temperature at the ICP–MS interface [5]. 2D modelled plasma temperature profiles at (a) 750, (b) 1000, (c) 1250 and (d) 1500 W. These were calculated for a 1 l min^{-1} injector gas flow rate, an 0.4 l min^{-1} auxiliary gas flow rate and a 12 l min^{-1} outer gas flow rate. The diameter of the torch injector tube inlet was 1.5 mm while the sampling cone orifice had a diameter of 1 mm. [*Note*: the small black lines above each panel represent the extent of the length of the cool central channel.] Reproduced with permission of the RSC.

heavier it is simpler to leave it on the horizontal!). Because of this horizontal positioning of the ICP torch in relation to the spectrometer, all species that enter the plasma are transferred into the mass spectrometer. All the associated sample-introduction devices (see Chapter 4) and method of operation of the ICP (see Section 5.3) are the same. As indicated in Section 5.3, the temperature of the ICP is not in local thermodynamic equilibrium. Attempts at modelling the temperature, at various applied powers have been reported (Figure 7.4) [5]. In these theoretical studies it is interesting to note that the increasing applied power (from 750 to 1500 W) creates a larger area of hotter temperature closer to the sampling orifice of the ICP–MS interface. Conversely, the cooler central channel is more elongated at lower applied power.

7.3.2   The Interface

The main feature in the success of this technique has been the development of a suitable interface. The interface allows the coupling of the atmospheric ICP source with the high-vacuum mass spectrometer, while still maintaining a high degree of sensitivity. The interface consists of a water-cooled outer sampling cone that is positioned near the plasma source (Figure 7.5). The sampling cone is typically made of nickel because of its high thermal conductivity, relative

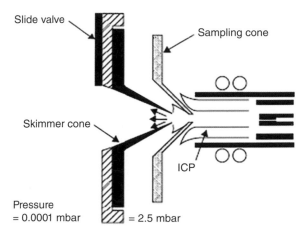

Slide valve

Sampling cone

Skimmer cone

ICP

Pressure
= 0.0001 mbar   = 2.5 mbar

Figure 7.5 Schematic diagram of the inductively coupled plasma–mass spectrometer interface.

resistance to corrosion and its robust nature. The pressure differential created by the sampling cone is such that ions from the plasma and the plasma gas itself are drawn into the region of lower pressure through the small orifice of the cone (≈ 1.0 mm). The region behind the sampling cone is maintained at a moderate pressure (≈ 2.5 mbar) by using a rotary vacuum pump. As the gas flow through the sample cone is large, a second cone is placed close enough behind the sampling cone to allow the central portion of the expanding jet of plasma gas and ions to pass through the skimmer cone. The latter (typically made of nickel) has an orifice diameter of ≈ 0.75 mm. The pressure behind the skimmer cone is maintained at $\approx 10^{-4}$ mbar. The extracted ions are then focused by a series of electrostatic lenses into the mass spectrometer. If the ICP–MS system contains a collision/reaction cell (see Section 7.6.3), then this is located before the mass spectrometer.

7.4   Mass Spectrometers

The mass spectrometer acts as a filter, transmitting ions with a pre-selected mass/charge ratio. The transmitted ions are then detected and converted into an appropriate form for display. A range of mass spectrometers have been coupled to an ICP. However, the first to be exploited commercially was the quadrupole mass spectrometer. This was closely followed by the high-resolution mass spectrometer to overcome some of the deficiencies of a quadrupole mass spectrometer, that is, its inability to overcome interferences (see Section 7.6) due to its low resolution; a quadrupole mass spectrometer being limited to unit-mass resolution. More recently, other mass spectrometers have been exploited, namely the triple quadrupole, ion-trap, time-of-flight (TOF) and multiple collector mass spectrometers.

7.4.1 Quadrupole Mass Spectrometer

The quadrupole analyser consists of four straight metal rods positioned parallel to and equidistant from the central axis (Figure 7.6). By applying direct current (DC) and radiofrequency (RF) voltages to opposite pairs of the rods, it is possible to have a situation where the DC voltage is positive for one pair and negative for the other. Likewise, the RF voltages on each pair are 180° out of phase, that is, they are opposite in sign, but with the same amplitude. Ions entering the quadrupole are subjected to oscillatory paths by the RF voltage. However, by selecting appropriate RF and DC voltages, only ions of a given mass/charge ratio will be able to traverse the length of the rods and emerge at the other end. Other ions are lost within the quadrupole analyser; as their oscillatory paths are too large, they collide with the rods and become neutralized.

ICP–MS can be operated in two distinctly different modes; that is, with the mass filter transmitting only one mass/charge ratio or with the DC and RF values being changed continuously. The former would allow single-ion monitoring, with the latter allowing multi-element analysis. In single-ion monitoring, all the data is obtained from a single mass/charge ratio; although this precludes the major facet of the technique, it does provide more sensitivity for the element (mass/charge ratio) of interest. For multi-element analysis, RF and

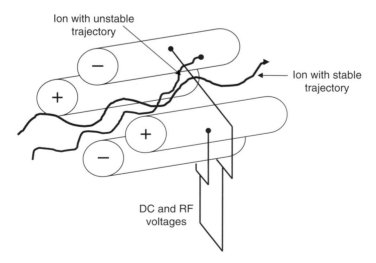

Figure 7.6 Schematic arrangement of the quadrupole analyser arrangement.

DC voltage scanning is required. Scanning, and hence data acquisition, can be carried out in two different modes, as follows:

- peak jumping or peak hopping
- (multi-channel) scanning

These different scanning modes are illustrated in Figure 7.7, using Tl, Pb and Bi as an example. Thallium has isotopes at: ^{203}Tl (which is 29.5% abundant) and ^{205}Tl (70.5%); Lead has isotopes at: ^{204}Pb (which is 1.4% abundant), ^{206}Pb (which is 24.1%), ^{207}Pb (which is 22.1%) and ^{208}Pb (which is 52.4%) and Bismuth, which is 100% abundant at ^{209}Bi. In peak hopping or peak jumping mode, the signal ions are measured at selected mass/charge ratios for a 'dwell' time (e.g. 0.5–1.0 s). This allows fast repetitive analyses of a pre-determined set of elements. However, it does not allow interrogation of the mass spectrum for potential interferences, for example, unexpected polyatomic interferences (see Section 7.6.2). In multi-channel scanning, all mass/charge ratios

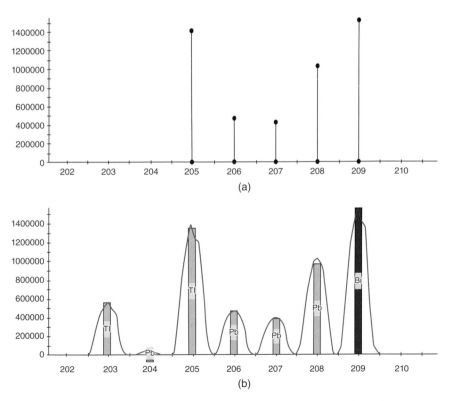

Figure 7.7 Different methods of data acquisition (scanning modes) employed for inductively coupled plasma–mass spectrometry, using Tl, Pb and Bi as an example: (a) peak jumping (or hopping) and (b) (multi-channel) scanning mode.

are repetitively scanned (e.g. between 45 and 210), thus providing a complete 'fingerprint' of the unknown sample composition. This mode can also be useful for qualitative analysis of unknown samples. In the multi-channel scanning mode, the typical dwell time per mass/charge ratio is 0.1–0.5 ms. The mass analyser is therefore a very rapid sequential spectrometer.

Quadrupole mass analysers are capable of only unit-mass resolution; that is, they can observe integral values of the mass/charge ratio only (e.g. 204, 205, 206 etc.). An important criterion in ICP–MS is the ability to measure a weak signal intensity at mass, M, from an adjacent major peak; that is, $M + 1$ or $M - 1$ (see Figure 7.1). This is termed abundance sensitivity. As you might imagine, this is an important factor in ICP–MS, where ultra-trace analysis is being carried out against a background of major element impurities (or the sample matrix). Values of up to 10^6 are achievable in quadrupole ICP–MS instruments.

7.4.2 High-Resolution Mass Spectrometers

The lack of resolution for a quadrupole instrument (i.e. limited to unit-mass resolution) makes it impossible to separate interferences (see Section 7.6) that coincide at the same nominal mass. A spectral resolution of ≈10 000 would remove most of the interferences from polyatomic species, but much higher resolution is required for isobaric interferences (Table 7.2). To separate at high-resolution, a mass spectrometer capable of much higher resolution is required. A range of high-resolution mass spectrometers have been deployed and include tandem MS (triple quadrupole ICP or ICP–QQQ), sector-field (SF) instruments based on double-focusing (DF) with magnetic and electric field separation (i.e. SF–ICP–MS or DF–ICP–MS) and multiple collector (MC–ICP–MS) (Figure 7.8). While the use of an ICP with a high-resolution mass spectrometer was first reported in 1989 [6, 7]. A high-resolution sector-field mass analyser (Figure 7.8a) consists of an electrostatic analyser (ESA) and a magnetic analyser. In addition to both analysers, the important aspect of the high-resolution instrument is the use of narrow entry and exit slits that control the number of ions passing through to the detector at any one time. This is known as a high-resolution mass analyser as it can focus both the energy and m/z ratio. Ions from the ICP pass through a narrow slit, hence allowing only ions correctly aligned on an axial plane to pass through, so resulting in a narrow beam of ions all travelling parallel to each other. The ESA consists of two curved plates applied with a DC voltage, thus allowing the inner plate (negative polarity) to attract positively charged ions, while the outer plate (positive polarity) repels the ions. The ion beam then passes between the two plates and is both focused and curved through an angle of 40°. Since only ions with a narrow range of kinetic energies (see Eq. (7.2)) can pass through the ESA, the latter thus forms an effective energy 'filter', so allowing ions of all masses to pass

Table 7.2 Examples of the resolution required to separate ions of similar intensity. Reproduced with permission of Elsevier.

Analyte ion	Interfering ion	Resolution[a]
$^{24}Mg^+$	$^{12}C_2^+$	1605
$^{28}Si^+$	$^{14}N_2^+$	958
$^{28}Si^+$	$^{12}C^{16}O^+$	1557
$^{44}Ca^+$	$^{12}C^{16}O_2^+$	1281
$^{51}V^+$	$^{35}Cl^{16}O^+$	2572
$^{52}Cr^+$	$^{40}Ar^{12}C^+$	2375
$^{54}Fe^+$	$^{40}Ar^{14}N^+$	2088
$^{56}Fe^+$	$^{40}Ar^{16}O^+$	2502
$^{63}Cu^+$	$^{40}Ar^{23}Na^+$	2790
$^{64}Zn^+$	$^{32}S^{16}O_2^+$	1952
$^{75}As^+$	$^{40}Ar^{35}Cl^+$	7775
$^{80}Se^+$	$^{40}Ar_2^+$	9688

Source: Adapted from reference [14].
a)   Resolution = $M/\Delta M$.

into the magnet analyser. In the energy field of the magnet, the ions are separated by their m/z ratios, such that ions of different masses follow different circular trajectories. By setting the field strength of the magnet, it is possible to select only those ions with a specific m/z ratio. The ion beam then passes through a narrow slit situated at the focal point of the magnet (collector slit). However, the high cost of these instruments precludes their wide availability. Figure 7.9 shows the resolution achievable by using this type of mass spectrometer for (a) the separation of iron from ArO^+ and (b) the determination of silicon in a steel sample. A high-resolution ICP–MS (HR–ICP–MS), sector-field ICP–MS (SF–ICP–MS) or double-focusing ICP–MS (DF–ICP–MS) (e.g. a reverse geometry double-focusing magnetic sector MS) is therefore used to eliminate polyatomic interferences (see Section 7.6.2) due to its higher resolution (i.e. a capability to resolve differences in m/z to 0.01–0.001 amu, compared to a quadrupole-based system that can resolve to ~0.5–1 amu).

In addition, other mass spectrometers have been used with an ICP to alleviate or remedy specific issues. For example, an ICP–tandem mass spectrometer (triple quadrupole ICP–MS/MS or ICP–QQQ) (Figure 7.8b) is used to eliminate molecular interferences (see Section 7.6.3), based on its inclusion of a collision/reaction cell (see Section 7.6.3). Finally, a multiple collector ICP–MS (MC–ICP–MS) (Figure 7.8c) is used to significantly improve isotope ratio measurements of long-lived, extinct radiogenic isotope systems or stable isotopes [8]. The main feature of the instrument that allows this is the capability for simultaneous detection of isotopes by use of multiple detectors. For example, isotope precision with a quadrupole-based ICP–MS instrument is limited to ≥0.1% RSD, with the inclusion of a collision/reaction cell the same system can achieve a precision as low as ≤0.05% RSD. This is comparable to the precision achievable with both sector-field ICP–MS and

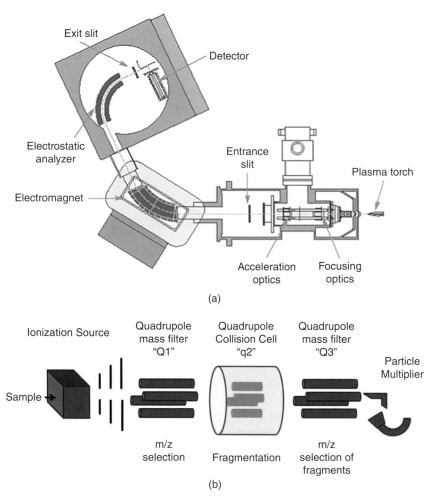

Figure 7.8 Schematic diagrams of the layout of high-resolution mass spectrometers for inductively coupled plasma (a) high-resolution ICP–MS (HR-ICP–MS), sector-field ICP–MS (SF-ICP–MS) or double-focusing ICP–MS (DF-ICP–MS) (e.g. a reverse geometry double-focusing magnetic sector MS) [7] (Reproduced with permission of the RSC), (b) ICP–tandem mass spectrometer (triple quadrupole ICP–MS or ICP–QQQ) and (c) multiple collector ICP–MS (MC–ICP–MS) [8] (Reproduced with permission of the RSC).

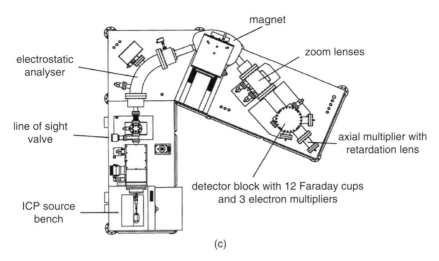

(c)

Figure 7.8 (*Continued*)

ICP–ToF–MS instruments (see Section 7.4.4) (≤0.05% RSD), whereas with a MC–ICP–MS the precision can be as good as 0.002% RSD [9].

The ICP–MS/MS (or triple quadrupole ICP–QQQ) instrument (Figure 7.8b) operates in two distinct modes: single (or 'open') quadrupole, SQ, mode or double mass selection, MS/MS, mode (Figure 7.10a) [10]. In the case of the SQ mode (Figure 7.10b) Q1 (quadrupole 1) can be operated as an ion guide only or as a bandpass mass filter. In this first case (Figure 7.10b), Q1 is 'fully open', so all ions, that is, the analyte ($^{m1}A^+$), the interfering ions with the same m/z ($^{m1}I^+$), and concomitant ions with different m/z ($^{m2}C^+$) enter the Collision Reaction Cell (CRC). [*Note*: This situation is like the events with a conventional ICP–CRC–(quadrupole)MS system, and can be used in situations that do not require significant interference removal, where no interferences occur and in simple sample matrices.] In the MS/MS mode, however (Figure 7.10b), both Q1 and Q2 act as real mass filters, with both quadrupoles selecting the same m/z in the on-mass approach or different m/z's in the mass-shift approach, depending on whether the interfering or the analyte ion reacts with the reagent gas added. In MS/MS mode (Figure 7.10b), a specific m/z (that of the target ion, $^{m1}A^+$) is selected in Q1, such that only the analyte ion and the interfering ions with the same m/z ($^{m1}I^+$), can enter the CRC. All concomitant ions with different m/z ($^{m2}C^+$) are efficiently removed by Q1. Operating in MS/MS mode leads to a lower sensitivity due to the reduced ion transmission efficiency, however, several advantages are evident by operating in this mode. The advantages of operating in MS/MS mode leads to avoidance of unwanted product ions resulting from reactions between other ions and the reaction gas, a reduction in non-spectral interferences or matrix effects and an improvement in the

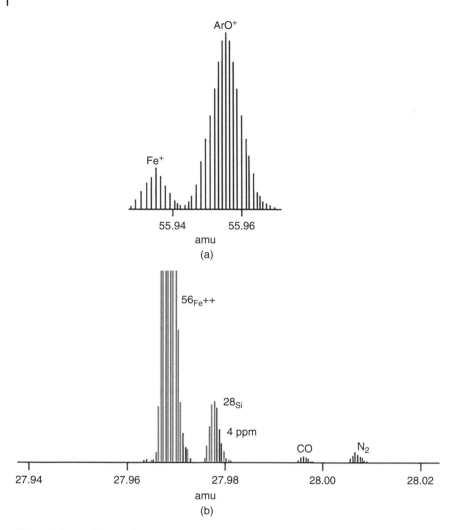

Figure 7.9 Resolution achievable using a sector-field mass spectrometer for (a) the separation of iron from ArO$^+$ and (b) the determination of silicon in a sample of steel.

efficiency of ion-molecule chemistry occurring in the cell. This allows the ICP–MS/MS to deliver interference-free ultra-trace element determination by selection of appropriate reaction gases and identifying the best reaction product ion for monitoring.

In terms of ICP–MS/MS, different scanning options are also possible that facilitate interference-free isotopic determination and include product ion scan, precursor ion scan and neutral mass gain/loss scan (Figure 7.11) [10]. The product ion scan is used to identify the best suited reaction product ion

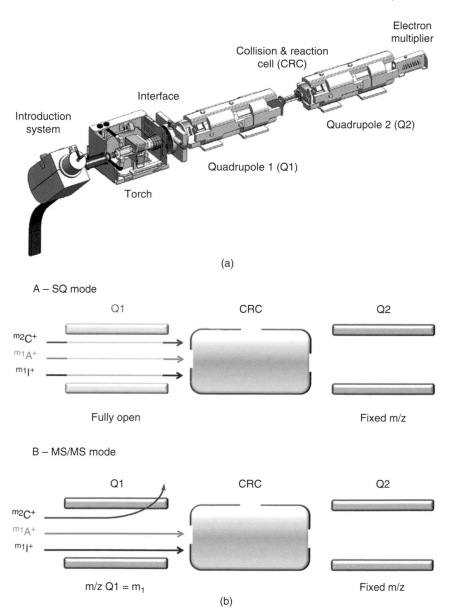

(a)

A – SQ mode

(b)

Figure 7.10 Schematic representation of (a) ICP–tandem mass spectrometer (triple quadrupole ICP–MS or ICP–QQQ) and (b) its operating modes [10]. Reproduced with permission of the RSC.

from the reaction between the original analyte ion and the cell reaction gas for interference-free determination with the highest signal-to-background ratio. To perform this scan, Q1 should be fixed at a specific nominal m/z ratio ($^{m1}A^{+}$) (Figure 7.11a). By fixing the m/z ratio of $m1$ and pressurizing the cell with the reaction gas (R), leads to the formation of different reaction product ions (e.g. $^{m1}A\,^{mR}R_x^{+}$). All the formed reaction product ions can be observed by scanning Q2 (i.e. between 2 and 260 amu). From the resultant product ion spectrum, the most appropriate reaction product ion can be selected for subsequent method development. In the precursor ion scan (Figure 7.11b) more information can be obtained about the origin of a specific product ion that has resulted from the reaction between a so-far unidentified interference ion and the reaction gas (R), which could hinder interference-free determination of the analyte ion or a reaction product ion. Using precursor ion scanning, the m/z of the product ion

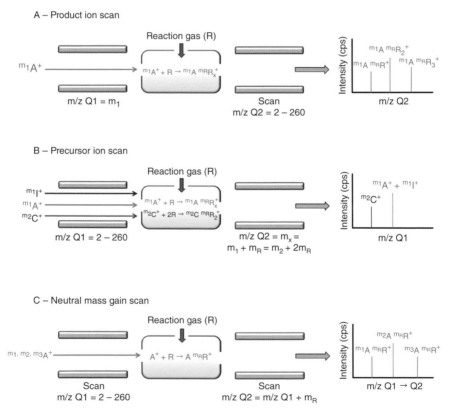

Figure 7.11 Schematic representation of the operating principles for the different scanning options available for a ICP–tandem mass spectrometer (triple quadrupole ICP–MS or ICP–QQQ). (a) Product ion scan, (b) precursor ion scan and (c) neutral mass gain scan [10]. Reproduced with permission of the RSC.

selected is fixed by Q2, in the presence of the reaction gas, producing a precursor ion spectrum (2–260 amu) with Q1. From the resultant spectrum, precursor ions giving rise to spectral overlap at the m/z of interest, for the analyte, can be identified. Finally, in neutral mass gain/loss scan mode, both quadrupoles are used in scanning mode with a fixed constant difference in m/z between Q1 and Q2; this fixed constant difference, K = m/z(Q2) – m/z(Q1), where K > 0 in neutral mass gain scan mode and K < 0 in a neutral mass loss scan. This allows both gain or loss of mass to be investigated, depending on whether the precursor ion is lower or higher in mass than the corresponding product ion. The main application of neutral mass gain in ICP–MS/MS (Figure 7.11c) is the study of the isotopic pattern of the target analyte (A). For example, an element with three isotopes ($^{m1,m2,m3}A^+$) and a reagent gas (R), reacts with the original analyte ions as follows:

$$^{m1,m2,m3}A^+ + {}^{mR}R^+ \rightarrow {}^{m1,m2,m3}A^{mR}R^+ \tag{7.2}$$

In this case, the neutral gain scan can be used to evaluate whether the reaction product ions $^{m1,m2,m3}A^{mR}R^+$ show the same isotopic pattern as does the corresponding precursor ions. Therefore, Q1 is selected at m/z = $m1$, $m2$ and $m3$, and Q2 at m/z $m1 + mR$, $m2 + mR$ and $m3 + mR$, respectively, such that K = mR during the entire scanning process. The use of the neutral mass gain spectrum is therefore a useful diagnostic tool in the context of isotopic analysis.

The principle of overcoming spectral interferences for ICP–MS/MS is illustrated in Figure 7.12 [10]. Two options are possible: the on-mass and the mass-shift approaches. In the on-mass approach (also known as the direct determination), both Q1 and Q2 are set at the same m/z; that is, $^{m1}A^+$. The reaction between the reaction gas (R) and the interfering ion ($^{m1}I^+$) is used to remove the contribution to the signal intensity at the measured m/z. The newly formed reaction product ion ($^{m1}I^{mR}R^+$) is removed by means of Q2, while the original analyte ion is measured at the original m/z (Figure 7.12a). [$Note$: this is the normal method used in ICP–CRC–MS.] In the mass-shift approach, Q1 is fixed at the original m/z of the analyte of interest ($^{m1}A^+$), such that only interfering ions with the same m/z ($^{m1}I^+$) can reach the CRC, while all other ions with different m/z's ($^{m1}C^+$) are removed. In this case, the CRC is pressurized with a reaction gas that has either a high reactivity towards the analyte ion and no reactivity towards the interfering ion (Figure 7.12b) or a different behaviour towards the interfering ion (Figure 7.12c). If several type of reaction product ions are formed, the best suited reaction is selected by fixing the m/z of Q2. In this case therefore ions that interfere with the analyte ion at its original m/z will enter the cell, but are ejected by Q2 when they show no, or a different, reactivity towards the reaction gas. An example of the use of on-mass, with H_2 and the mass-shift method, with O_2, for overcoming spectral interferences for ^{80}Se is

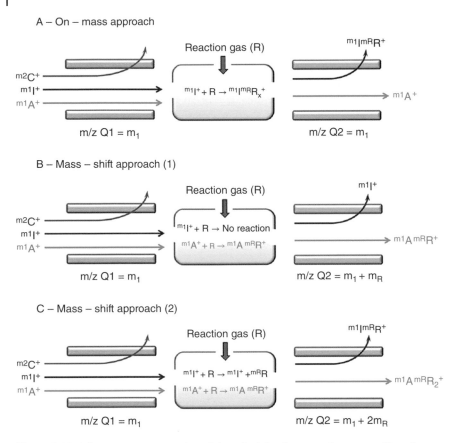

Figure 7.12 Schematic representation of the principle of overcoming spectral interferences for a ICP–tandem mass spectrometer (triple quadrupole ICP–MS or ICP–QQQ). (a) On-mass approach, (b) mass-shift approach (1) and (c) mass-shift approach (2) [10]. Reproduced with permission of the RSC.

shown in Figure 7.13 [10]. The signal for selenium isotopes is affected by the argon-based polyatomic ions; that is, $^{40}Ar^{40}Ar^+$ on $^{80}Se^+$ (Table 7.4). In the on-mass approach Figure 7.13a, the addition of H_2 as the reaction gas leads to formation of $^{40}Ar^1H^+$ ($^{40}Ar^{40}Ar^+ + H_2 \rightarrow {}^{40}Ar^1H^+ + Ar + H$), whereas the addition of O_2, in the mass-shift approach in Figure 7.13b, allows formation of $^{80}Se^{16}O^+$ (at m/z 96 amu).

7.4.3 Ion-Trap Mass Spectrometer

An ion-trap mass spectrometer consists of a cylindrical ring electrode and two end-cap electrodes (Figure 7.14). The top end-cap allows ions to be introduced into the trap where they are retained (i.e. ion-trapping). An RF voltage

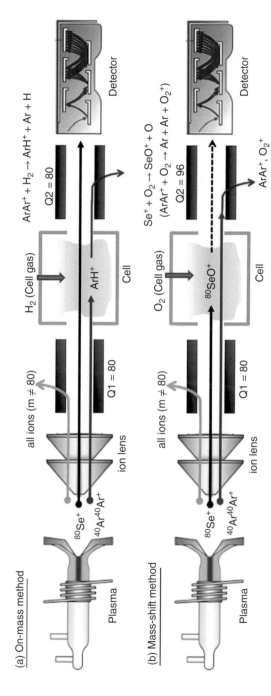

Figure 7.13 An example of the principle of overcoming spectral interferences for a ICP–tandem mass spectrometer (triple quadrupole ICP–MS or ICP–QQQ) for the spectral free determination of ^{80}Se operated in MS/MS mode using both the (a) on-mass approach and (b) mass-shift approach [10]. Reproduced with permission of the RSC.

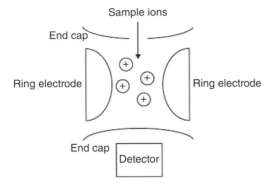

Figure 7.14 Schematic representation of the operation of an ion-trap mass spectrometer.

is applied to the ring electrode to stabilize and retain ions of different m/z ratios. Then, by increasing the applied voltage the paths of the ions of successive m/z ratios are rendered unstable. These ions then exit the trap via the bottom end-cap, prior to detection. Such a process can be described as 'being in a mass-selective instability mode'.

7.4.4 Time-of-Flight Mass Spectrometer

A TOF mass spectrometer does not rely on magnetic, electrostatic, or an RF field to separate ions of different m/z ratios, but instead uses an applied accelerating voltage and the resultant different velocities of ions as the basis of its separation. As each ion of different m/z ratios has the same kinetic energy (see Eq. (7.3)) but a different mass, it will travel through the TOF spectrometer at a different velocity and as a result be separated. The kinetic energy (KE) is defined as follows:

$$KE = mv^2 / 2 \qquad\qquad (7.3)$$

where m is the mass of the ion and v its velocity.

As a result, the TOF mass spectrometer measures the time it takes for an ion to travel through the 'drift tube' from the source to the detector (Figure 7.15). The lighter ions, which are travelling faster, reach the detector before the heavier ions. To increase the drift path length, an ion reflector, or 'reflectron', is introduced into the mass analyser – this reverses the direction of flow of the ions, as well as doubling the flight path.

The disadvantage of the TOF mass spectrometer is that it requires a pulsed source of ions, whereas an ICP provides a continuous ion beam. One approach to circumnavigate this issue is by the addition of a quadrupole mass analyser (see Section 7.4.1) prior to the TOF mass spectrometer.

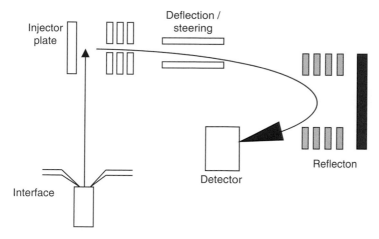

Figure 7.15 Schematic diagram of the layout of a time-of-flight mass spectrometer.

An overview of the diverse range of mass spectrometers coupled with an ICP is shown in Figure 7.16. Most are commercially available, but certainly all have been applied to different general and niche applications.

7.5 Detectors

Three types of detectors are used in mass spectrometry (Figure 7.17): the discrete-dynode electron multiplier (Figure 7.17a), continuous-dynode electron multiplier (also known as the channel electron multiplier) (Figure 7.17b) and the Faraday cup (Figure 7.17c). The electron multiplier was first invented in 1935 [11]. Both electron multipliers operate via the same basic principle; that is, secondary electron emission. Essentially, when a charged particle (e.g. an ion or electron) strikes a surface it causes secondary electrons to be released from atoms in the surface layer. A Faraday cup (Figure 7.17c), on the other hand, captures charged particles (e.g. ions) under vacuum conditions resulting in a change in electrical current.

The most common type of detector is the continuous-dynode electron multiplier (Figure 7.17a). The operating principles of this electron multiplier are like those of the photomultiplier tube (see Section 6.5.1), apart from the absence of dynodes. In addition, the electron multiplier must operate under vacuum conditions ($<5 \times 10^{-5}$ Torr). This device consists of an open tube with a wide entrance cone, with the inside of the tube being coated with a lead oxide semiconducting material. The cone is biased with a high negative potential (e.g. $-3\,\mathrm{kV}$) at the entrance and held 'at ground' near the collector.

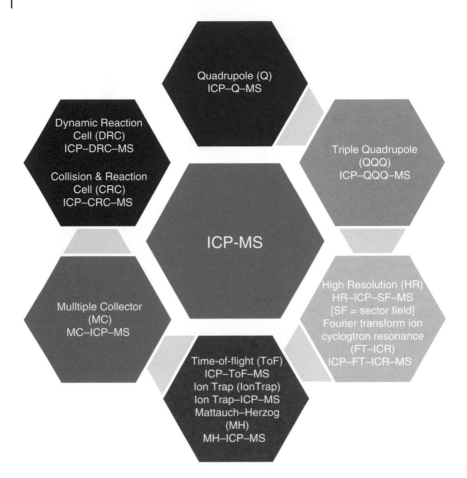

Figure 7.16 Overview of ICP–MS technologies.

Any incoming positive ions, from the mass analyser, are attracted towards the negative potential of the device. On impact, the positive ions cause one or more (secondary) electrons to be ejected. These secondary electrons are attracted towards the grounded collector within the electron multiplier. In addition, the initial secondary electrons can also collide with the surface coating, so causing further electrons to be ejected. This multiplication of electrons continues until all the electrons (up to 10^8 electrons) are collected. Such a discrete pulse of electrons is further amplified exterior to the electron multiplier and recorded as several ion 'counts per second'. All electron multiplier tubes have a limited lifetime, determined by the total accumulated charge, which is monitored.

The electron multiplier can also respond to photons of light from the ICP. For this reason, the detector can be mounted 'off-axis', the spectrometer can be

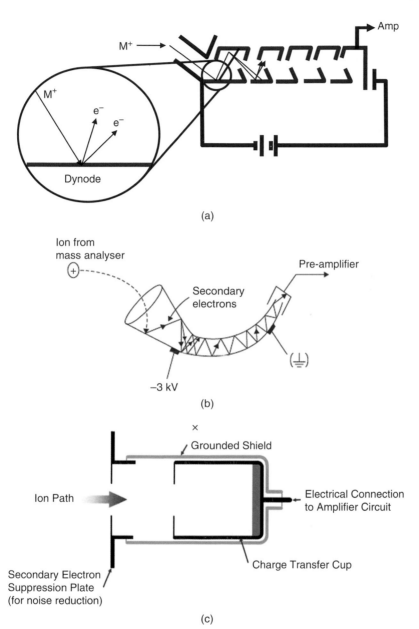

Figure 7.17 Detectors for mass spectrometry: (a) a discrete-dynode electron multiplier tube (EMT): mode of operation, (b) a continuous dynode (or channel) EMT: mode of operation and (c) Faraday cup detector: mode of its operation. *Source:* Reproduced with permission of ThermoFisher Scientific. https://www.thermofisher.com/uk/en/home/industrial/spectroscopy-elemental-isotope-analysis/spectroscopy-elemental-isotope-analysis-learning-center/trace-elemental-analysis-tea-information/inductively-coupled-plasma-mass-spectrometry-icp-ms-information/icp-ms-systems-technologies.html.

'off-axis' or a baffle can be in the centre of the ion lens. Nevertheless, the electron multiplier tube is a sensitive detector for ICP–MS.

An alternative detector that can be used when ion currents exceed 10^6 ions s^{-1} is the Faraday cup (Figure 7.17c). Ions impinging inside this detector are further amplified and counted. The advantage of this type of detector is its stability and freedom from 'mass bias', so allowing it to record highly accurate isotope ratio measurements. Its major disadvantage is its lack of sensitivity for trace elemental analysis. Another detector used in mass spectrometry is the 'Daly detector'. This consists of a metal knob that emits secondary electrons when struck by an ion. The secondary (generated) electrons are accelerated onto a scintillator material that produces light, this can then be detected using a photomultiplier tube (see Section 6.5.1).

7.6 Interferences

While interferences in ICP–MS are not so prevalent as in ICP–AES, nevertheless, some types of interferences do occur. The types of interferences can be broadly classified, according to their origin, into spectral (isobaric and molecular) and non-spectral. Spectral interferences can occur because of the overlap of atomic masses of different elements; that is, isobaric interferences and molecular processes. The latter can occur because of the acid(s) used to prepare the sample and/or the argon plasma gas (polyatomics). In addition, the formation of oxides, hydroxides and doubly charged species is possible. Non-spectral interferences or matrix interferences result in signal enhancement or depression with respect to the atomic mass.

7.6.1 Isobaric Interferences

These types of interferences are well characterized (Table 7.3) and because ~70% of the elements in the Periodic Table have more than one isotope, can usually be avoided by selecting an alternative isotope. By considering Table 7.3, it is possible to identify situations in which potential problems are alleviated by the ability to select an alternative isotope.

For example, if you were going to analyse nickel in a stainless steel sample, it would be appropriate to select the most abundant isotope for nickel, which occurs at atomic mass 58 and is 67.9% abundant. However, iron (0.31% abundant) occurs at the same mass. Thus, to prevent isobaric interferences, it may be necessary to select an alternative mass. At atomic mass 60, nickel is 26.2% abundant and so no interfering isotopes occur. One other point to consider is that by selecting a less-abundant isotope for nickel potentially leads to a lowering in sensitivity. This latter point may not be significant if nickel is present at a high enough concentration in the steel sample because of the inherent sensitivity of the technique.

Table 7.3 Isobaric interferences from Period 4 of the Periodic Table.

Atomic mass	Element of interest (% abundance)	Interfering element (% abundance)
39	K (93.10)	—
40	Ca (96.97)	Ar (99.6)[a]; K (0.01)
41	K (6.88)	—
42	Ca (0.64)	—
43	Ca (0.14)	—
44	Ca (2.06)	—
45	Sc (100)	—
46	—	Ca (0.003); Ti (7.93)
47	—	Ti (7.28)
48	Ti (73.94)	Ca (0.19)
49	—	Ti (5.51)
50	—	Ti (5.34); V (0.24); Cr (4.31)
51	V (99.76)	—
52	Cr (83.76)	—
53		Cr (9.55)
54	—	Cr (2.38); Fe (5.82)
55	Mn (100)	—
56	Fe (91.66)	—
57	—	Fe (2.19)
58	Ni (67.88)	Fe (0.33)
59	Co (100)	—
60	Ni (26.23)	—
61	—	Ni (1.19)
62	—	Ni (3.66)
63	Cu (69.09)	—
64	Zn (48.89)	Ni (1.08)
65	Cu (30.91)	—
66	Zn (27.81)	—
67	—	Zn (4.11)
68	—	Zn (18.57)
69	Ga (60.40)	—
70	—	Zn (0.62); Ge (20.52)

(*Continued*)

Table 7.3 (Continued)

Atomic mass	Element of interest (% abundance)	Interfering element (% abundance)
71	Ga (39.60)	—
72	Ge (27.43)	—
73	—	Ge (7.76)
74	Ge (36.54)	Se (0.87)
75	As (100)	—
76	—	Ge (7.76); Se (9.02)
77	—	Se (7.58)
78	Se (23.52)	Kr (0.35)
79	Br (50.54)	—
80	Se (49.82)	Kr (2.27)
81	Br (49.46)	—
82	—	Se (9.19); Kr (11.56)
83	Kr (11.55)	—
84	Kr (56.90)	Sr (0.56)[b]
85	—	—
86	—	Kr (17.37); Sr (9.86)[b]

a) Not in Period 4 of the Periodic Table but included because of its origin from the plasma source.
b) Not in Period 4 of the Periodic Table but included for completeness.

Unfortunately, other situations exist that do not have such readily amenable solutions. Probably the best example of this is the determination of calcium (atomic mass 40). Unfortunately (for Ca), the ICP source consists of argon ions that have a mass coincidence at atomic mass 40 (Ar, 99.6% abundant). In this situation, no alternative mass exists for Ca that will provide any degree of sensitivity. The best available mass is 44 amu at which Ca has an abundance of 2.08%. An alternative approach is to use collision/reaction cell technology (see Section 7.6.3).

7.6.2 Molecular Interferences

Molecular interferences derive from several origins and can be sub-divided into two different types; that is, polyatomics and doubly charged polyatomic interferences.

Polyatomic interferences are derived because of selected interactions between the element of interest and its associated aqueous solution, the plasma gas (Ar) or the types of acid(s) used in the preparation of the sample (Table 7.4). The type of acid-derived interferences considered in Table 7.4 are due to nitric, sulfuric, hydrochloric and phosphoric acids. It now becomes evident that the

Table 7.4 Potential polyatomic interferences derived from the element of interest and its associated aqueous solution, the plasma gas itself and the type of acid(s) used to digest or prepare the sample.

Atomic mass	Element of interest (% abundance)	Polyatomic interference
39	K (93.10)	$^{38}Ar^{1}H^{+}$
40	Ca (96.97)	$^{40}Ar^{+}$
41	K (6.88)	$^{40}Ar^{1}H^{+}$
42	Ca (0.64)	$^{40}Ar^{2}H^{+}$
43	Ca (0.14)	—
44	Ca (2.06)	$^{12}C^{16}O^{16}O^{+}$
45	Sc (100)	$^{12}C^{16}O^{16}O^{1}H^{+}$
46	—	$^{14}N^{16}O^{16}O^{+}$; $^{32}S^{14}N^{+}$
47	—	$^{31}P^{16}O^{+}$; $^{33}S^{14}N^{+}$
48	Ti (73.94)	$^{31}P^{16}O^{1}H^{+}$; $^{32}S^{16}O^{+}$; $^{34}S^{14}N^{+}$
49	—	$^{32}S^{16}O^{1}H^{+}$; $^{33}S^{16}O^{+}$; $^{14}N^{35}Cl^{+}$
50	—	$^{34}S^{16}O^{+}$; $^{36}Ar^{14}N^{+}$
51	V (99.76)	$^{35}Cl^{16}O^{+}$; $^{34}S^{16}O^{1}H^{+}$; $^{14}N^{37}Cl^{+}$; $^{35}Cl^{16}O^{+}$
52	Cr (83.76)	$^{40}Ar^{12}C^{+}$; $^{36}Ar^{16}O^{+}$; $^{36}S^{16}O^{+}$; $^{35}Cl^{16}O^{1}H^{+}$
53	—	$^{37}Cl^{16}O^{+}$
54	—	$^{40}Ar^{14}N^{+}$; $^{37}Cl^{16}O^{1}H^{+}$
55	Mn (100)	$^{40}Ar^{14}N^{1}H^{+}$
56	Fe (91.66)	$^{40}Ar^{16}O^{+}$
57	—	$^{40}Ar^{16}O^{1}H^{+}$
58	Ni (67.88)	—
59	Co (100)	—
60	Ni (26.23)	—
61	—	—
62	—	—
63	Cu (69.09)	$^{31}P^{16}O_{2}{}^{+}$
64	Zn (48.89)	$^{31}P^{16}O_{2}{}^{1}H^{+}$;$^{32}S^{16}O^{16}O^{+}$;$^{32}S^{32}S^{+}$
65	Cu (30.91)	$^{33}S^{16}O^{16}O^{+}$; $^{32}S^{33}S^{+}$
66	Zn (27.81)	$^{34}S^{16}O^{16}O^{+}$; $^{32}S^{34}S^{+}$
67	—	$^{35}Cl^{16}O^{16}O^{+}$
68	—	$^{40}Ar^{14}N^{14}N^{+}$; $^{36}S^{16}O^{16}O^{+}$; $^{32}S^{36}S^{+}$
69	Ga (60.40)	$^{37}Cl^{16}O^{16}O^{+}$

(Continued)

Table 7.4 (Continued)

Atomic mass	Element of interest (% abundance)	Polyatomic interference
70	—	$^{35}Cl_2^+$; $^{40}Ar^{14}N^{16}O^+$
71	Ga (39.60)	$^{40}Ar^{31}P^+$; $^{36}Ar^{35}Cl^+$
72	Ge (27.43)	$^{37}Cl^{35}Cl^+$; $^{36}Ar^{36}Ar^+$; $^{40}Ar^{32}S^+$
73	—	$^{40}Ar^{33}S^+$; $^{36}Ar^{37}Cl^+$
74	Ge (36.54)	$^{37}Cl^{37}Cl^+$; $^{36}Ar^{38}Ar^+$; $^{40}Ar^{34}S^+$
75	As (100)	$^{40}Ar^{35}Cl^+$
76	—	$^{40}Ar^{36}Ar^+$; $^{40}Ar^{36}S^+$
77	—	$^{40}Ar^{37}Cl^+$; $^{36}Ar^{40}Ar^1H^+$
78	Se (23.52)	$^{40}Ar^{38}Ar^+$
79	Br (50.54)	$^{40}Ar^{38}Ar^1H^+$
80	Se (49.82)	$^{40}Ar^{40}Ar^+$
81	Br (49.46)	$^{40}Ar^{40}Ar^1H^+$
82	—	$^{40}Ar^{40}Ar^1H^1H^+$
83	Kr (11.55)	—
84	Kr (56.90)	—
85	—	—
86	—	—

spectra obtained from a quadrupole MS instrument may be more complicated than anticipated. Very few elements in Period 4 of the Periodic Table are unaffected by some type of interference (either isobaric or polyatomic). It is likely that the polyatomic ions are formed within the interface where the ions are undergoing transfer from the atmospheric source to the mass spectrometer vacuum. Nevertheless, they provide an unwelcome addition to the interpretation of mass spectra. Therefore, the unsuspecting analyst may record a signal due to the presence of a polyatomic interference rather than the isotope (element) intended.

In addition to the previously described polyatomic interferences derived essentially from the Ar plasma gas and the acids used, a further type of interference can be identified; doubly charged polyatomic interferences. This is due to the formation of doubly charged species. Remember that for an isotope (element), you are measuring its mass/charge ratio. If the charge, z, alters (normally $z = 1$), then the resultant mass/charge ratio will also change; for example, for a charge of two ($z = 2$), the resultant mass/charge ratio will halve. Of concern for this type of interference is the formation of doubly charged species of

Ce, La, Sr, Th and Ba. For example, for barium (atomic masses, 130, 132, 134, 135, 136, 137 and 138), at what atomic masses would the Ba^{2+} species occur? The Ba^{2+} ions would occur at half the mass of the parent singly charged ion. In order of importance due to the magnitude of their effect (greatest first) the Ba^{2+} ions would occur at 69, 68, 67, 66 and 65 atomic masses.

7.6.3 Remedies for Molecular Interferences

As a range of polyatomic interferences can and do occur in ICP–MS so remedies have been sought and found that allow these interferences to be reduced or minimized. The most important approach is the use of collision/reaction cells. Commercial systems are available with collision/reaction cells integral to the mass analyser.

While collision and reaction cells have been used extensively for fundamental studies in ion-molecule chemistry, it is only in this century that they have been applied to ICP–MS [12, 13]. The collision/reaction cell is normally located behind the sample/skimmer cone arrangement and before the mass analyser. The use of collision and reaction cells in ICP–MS allows for the following:

- neutralization of the most intense chemical ionization species
- interferent or analyte ion mass/charge ratio shifts

These processes are affected using a range of reaction types, including the following:

- charge transfer
- proton transfer
- hydrogen-atom transfer
- atom transfer
- adduct formation

General forms of each of these reaction types, with selected examples, are presented in the following (A, analyte; B, reagent; C, interferent):

7.6.3.1 Charge Transfer
The general form of charge exchange is:

$$C^+ + B \rightarrow B^+ + C \tag{7.4}$$

Charge exchange allows the potential removal of, for example, the argon plasma gas ion interference and the resultant formation of uncharged argon plasma gas, which is not then detected. An example, using ammonia, is:

$$Ar^+ + NH_3 \rightarrow NH_3{}^+ + Ar \tag{7.5}$$

7.6.3.2 Proton Transfer

The general form of atom transfer with a proton is:

$$CH^+ + B \rightarrow BH^+ + C \tag{7.6}$$

Proton transfer can remove the interference from, for example, ArH^+. This results in the formation of neutral (uncharged) argon plasma gas that is then not detected. An example, using hydrogen, is:

$$ArH^+ + H_2 \rightarrow H_3^+ + Ar \tag{7.7}$$

7.6.3.3 Hydrogen-Atom Transfer

The general form of atom transfer with a hydrogen atom is:

$$C^+ + BH \rightarrow CH^+ + B \tag{7.8}$$

Hydrogen-atom transfer can alleviate an interference by increasing the mass/charge ratio by one. An example, using hydrogen, is:

$$Ar^+ + H_2 \rightarrow ArH^+ + H \tag{7.9}$$

7.6.3.4 Atom Transfer

The general form of a condensation reaction is:

$$A^+ + BO \rightarrow AO^+ + B \tag{7.10}$$

The use of atom transfer, in common with adduct formation, can increase the mass/charge ratio. For example, creation of the oxide of the element will increase the mass/charge ratio by 16 amu (atomic weight of O = 16).

$$Ce^+ + N_2O \rightarrow CeO^+ + N_2 \tag{7.11}$$

7.6.3.5 Adduct Formation

The general form of adduct formation is:

$$A^+ + B \rightarrow AB^+ \tag{7.12}$$

Adduct formation, with, for example, ammonia, NH_3, allows the mass/charge ratio to increase by 17 amu (atomic weight of N = 14 and atomic weight of H = 1).

$$Ni^+ + NH_3 \rightarrow Ni^+.NH_3 \tag{7.13}$$

To illustrate the practical use of this approach involving collision/reaction cells, two specific case studies are shown:

Case Study 1

As indicated in Section 7.6.1, an important isobaric interference prevents the determination of calcium when using ICP–quadrupole MS (both calcium and argon have their major abundant ions at 40 amu; that is, 96.9 and 99.6% abundancies, respectively). This interference can be alleviated by using a charge transfer reaction, specifically:

$$Ca^+ + Ar^+ + H_2 \longrightarrow Ca^+ + Ar + H_2^+$$

Both calcium and argon ions coincide at *m/z* 40 amu

Argon atom formed which is NOT separated by the spectrometer (7.14)

The addition of hydrogen gas into the collision/reaction cell will remove any Ar ions present in the plasma but will not affect the Ca ions in the plasma. This approach can be used to determine calcium in an argon ICP. In general, however, the use of charge transfer to remove an interference will only work if the reagent gas to be used has an ionization potential between the interferent ion (e.g. Ar^+) and analyte ion (e.g. Ca^+). The other reaction types are not reliant on the ionization potential of the reagent gas but are dependent on thermodynamic and kinetic factors.

Case Study 2

The argument for the use of a high-resolution mass spectrometer (see Section 7.4.2) was demonstrated by its ability to separate iron from ArO^+ in Figure 7.9a (both result in signals at 56 amu). An alternative approach would be to use an atom transfer reaction that allows an analyte shift to occur:

$$Fe^+ + ArO^+ + N_2O \longrightarrow FeO^+ + ArO^+ + N_2$$

Both iron and argon oxide ions coincide at *m/z* 56 amu

Iron oxide ion now occurs at *m/z* 72 amu (7.15)

[*Note:* while the removal of this interference at 56 amu is beneficial, it also creates another interference at 72 amu. Germanium has an isotope at 72 amu, which is 27.5% abundant. However, the major germanium isotope is ^{74}Ge, which is 36.35% abundant.]

The use of collision/reaction cells can therefore lead to the 'chemical resolution' of isobaric and polyatomic interferences. However, it should also be noted that as they often result in different species being formed, this approach also has the potential to create other interferences. The major reagent gases used in ICP–MS [12] are:

- collision gas; for example, He
- charge transfer gases; for example, H_2, NH_3
- oxidation-reagent gases; for example, O_2, N_2O
- reduction-reagent gas; for example, H_2
- other reaction gas; for example, CH_4.

7.6.4   Non-Spectral Interferences: Matrix-Induced

Non-spectral interferences result from problems associated with the sample matrix. However, their origin within the system can vary. Still, the resultant effect of non-spectral interferences is a loss of sensitivity for elements.

Problems associated with the ICP: in the ICP, an incorrect choice of the appropriate sample-introduction device can lead to blockage problems in the nebulizer. Once this problem has been successfully negated, however, solids can still build-up on the sample cone of the ICP–MS interface. Both problems can lead to intermittent and erratic signal generation. Nevertheless, several remedies are possible and include the following:

- Choice of nebulizer; for example, the use of a high-solids nebulizer (see Section 4.2).
- Aqueous dilution of the sample matrix (which may lead to decreased sensitivity of the analyte of interest) or the use of flow injection (see Section 4.5). Alternatively, the intermittent introduction of a sample with a high-salts content, followed by washing in dilute acid.
- Use of an internal standard. Application of the latter (addition of an element not in the sample at a fixed concentration to the standards and samples) will compensate for fluctuations in signal response.
- Matrix-matched standards. Matrix-matching of samples with calibration solutions may also be advantageous.
- The method of standard additions (see Section 1.8).
- 'On-line' coupling of a chromatographic separation technique to ICP–MS (see Section 4.5). The chromatographic separation technique may allow the preferential separation of the matrix from the element of interest by an ion-exchange process or, if the element has been chemically attached to an organic molecule, by reversed-phase high performance liquid chromatography or gas chromatography.

Problems associated with the mass spectrometer: while mass-discrimination effects, which usually result in lower sensitivity, can be experimentally observed in the mass spectrometer, their exact mechanism is not known.

7.7 Isotope Dilution Analysis

The normal methods of calibration for atomic spectrometry are external calibration and the method of additions (see Section 1.8). However, the use of a mass spectrometer provides an alternative approach; that is, IDA. As any mass spectrometer is capable of measuring isotope ratios (ratios of the different isotopes of a single element), IDA uses this approach to provide a unique method of calibration. The essential feature of the IDA technique is that the element under investigation has more than one stable isotope; this applies to more than 70% of the elements in the Periodic Table. The basis of this approach is that isotope ratios before (the sample) and after 'spiking' (sample, plus spike) are measured. Then, by applying a mathematical solution, the concentration of the element in the sample can be determined. The main criterion for this approach is that the element spike must have an artificially enriched isotopic abundance; that is, the isotopic composition of the element is different from that which would normally occur. This is illustrated in Figure 7.18, where Figure 7.18a shows the natural occurring isotopic composition of lead, that is, lead that may be present in a sample, while Figure 7.18c shows the isotopic composition of an artificially enriched isotopic standard in which the abundance of ^{206}Pb has been increased by approximately 25%. The resultant mass spectrum will appear as shown in Figure 7.18(b). By measuring the isotopic ratio of the most abundant Pb isotope (^{208}Pb) with ^{206}Pb in the sample (208/206 Pb ratio $\approx$ 2.0) and comparing this with the resultant 208/206 Pb ratio in the sample plus spike (208/206 Pb ratio $\approx$ 1.0), an exact concentration of lead in the sample can be determined by using the following formula:

$$A = \left[x.B_2 \left(m_1 / m_2 \right) - B_1 \right] / \left(z - zx / y \right)$$

7.16

where A is the number of grams of element in the original sample, x the measured isotope ratio (^{208}Pb/^{206}Pb) in the spiked sample, B_1 the number of grams of ^{208}Pb in the enriched spike (i.e. the mass of total spike multiplied by the atom abundance of ^{208}Pb), B_2 the number of grams of ^{206}Pb in the enriched spike, that is, the mass of total spike multiplied by the atom abundance of ^{206}Pb, m_1 the atomic weight of ^{208}Pb, m_2 the atomic weight of ^{206}Pb, z the fractional abundance of ^{208}Pb and y the isotope ratio (^{208}Pb/^{206}Pb) in the original sample.

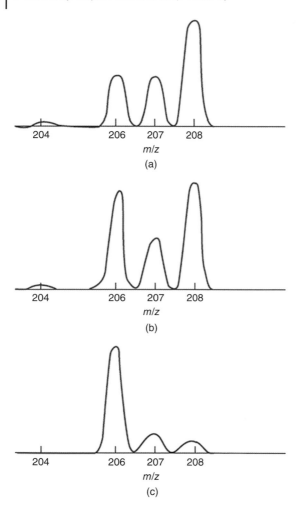

Figure 7.18 Mass spectra of isotopes of lead: (a) normal isotopic lead; (b) a mixture of normal isotopic lead, enriched with a ^{206}Pb 'spike' and (c) artificially enriched ^{206}Pb [5]. Reproduced with permission of the RSC.

It is worth noting that the values for B_1 and B_2 use the atom abundances for ^{208}Pb and ^{206}Pb, respectively, and not their percentage atom abundances as given in Table 7.5.

Table 7.5 Fractional abundance, by weight, of lead.

Isotope	203.973	205.974	206.976	207.977
Atom abundance (%)	1.425	24.144	22.083	52.347

Source: Data from NIST Certificate of Analysis SRM 981.

Case Study 3

Calculation of Fractional Abundance

Calculate the fractional abundance, by weight, of ^{208}Pb. The isotopic abundances of common lead are given in Table 7.5 (data from NIST Certificate of Analysis SRM 981). As the atomic weight of lead is 207.215, then the fractional abundance (by weight) of ^{208}Pb, z, can be calculated as follows:

$$z = (207.977 \times 52.347) / (203.973 \times 1.425) + (205.974 \times 24.144)$$
$$+ (206.976 \times 22.083) + (207.977 \times 52.347) = 0.5254 \qquad 7.17$$

It should be noted, however, that a quadrupole mass spectrometer may not measure the absolute isotopic composition of an element. It is likely, therefore, that the spectrometer will suffer from mass-discrimination effects. If this is the case, it will be necessary to apply corrections to the measured isotopic ratios before carrying out the calculation when using IDA.

Case Study 4

Example Calculation Using Isotope Dilution Analysis

If the concentration of lead in an aqueous sample (100 ml) was 100 ng ml^{-1}. Then the sample was 'spiked' with an enriched ^{206}Pb standard (3.5 µg) by the addition of a 5 ml volume. After ICP–MS analysis, the following isotope ratio results were obtained (Table 7.6). Data for the isotopic composition of common lead (from NIST Certificate of Analysis SRM 981) are in Table 7.5, while the corresponding data for the enriched ^{206}Pb 'spike' (from NIST Certificate of Analysis SRM 983) are in Table 7.7. By applying Eq. (7.16) confirm that the concentration of lead in the original sample was 100 ng ml^{-1}.

$$A = [(0.9521 \times (3.5 \times 10^{-6} \text{ g} \times 0.9215) \times (207.977 / 205.974)$$
$$- (3.5 \times 10^{-6} \text{ g} \times 0.0126)] / (0.5254 \quad 0.5254 \times 0.9521 / 2.1681)$$

$$A = [(0.9521 \times (3.225 \times 10^{-6} \text{ g})) \times (1.0097) - (4.410 \times 10^{-8} \text{ g})] / (0.2947)$$

$$A = [(3.100 \times 10^{-6} \text{ g}) - (4.410 \times 10^{-8} \text{ g})] / (0.2947)$$

$$A = [3.056 \times 10^{-6} \text{ g}] / (0.2947)$$

that is, 1.037×10^{-5} g in 105.0 ml of solution

or 9.876×10^{-8} g ml^{-1}

Table 7.6 Isotope ratio results for Case Study 4.

System	$^{208}Pb/^{206}Pb$
Original sample	2.1681
Sample plus enriched ^{206}Pb 'spike'	0.9521

Table 7.7 Isotopic composition of enriched ^{206}Pb 'spike'.

Isotope	203.973	205.974	206.976	207.977
Atom abundance (%)	0.0342	92.1497	6.5611	1.2550

Source: From NIST Certificate of Analysis SRM 983.

or $0.0988\,\mu g\,ml^{-1}$

or $98.8\,ng\,ml^{-1}$, which is very close to $100\,ng\,ml^{-1}$.

IDA can be applied to all elements with more than one isotope. In principle, therefore, it can be used for multi-element analysis. However, in practice it is normally reserved for single-element determinations due to the high cost of the enriched stable isotopes and the increased time of each analysis. However, IDA does provide an ideal approach to internal standardization, in which one of the element's own isotopes acts as the internal standard. Use of the latter is known to improve the precision of the data. It is for this reason that IDA–ICP–MS is most commonly used by agencies involved in the production and certification of reference materials or in studies relating to nutrition, bioavailability and speciation.

Case Study 5

Mass Spectral Interpretation

Mass spectral data obtained from an ICP–MS analysis are shown in Figure 7.19a–f. By using the data presented in Table 7.8, it is possible to interpret the mass spectra. For the purposes of the following exercise, you can assume no molecular interferences. However, remember that isobaric interferences can and do occur. The results are shown in Table 7.9 that show (a) the results for Ti, V, Cr, Mn, Fe, Ni, Cu and Zn, (b) Sr and Mo, (c) Ag, Cd and Ba, (d) Hf and Ta, (e) Tl, Pb and Bi and (f) Th and U.

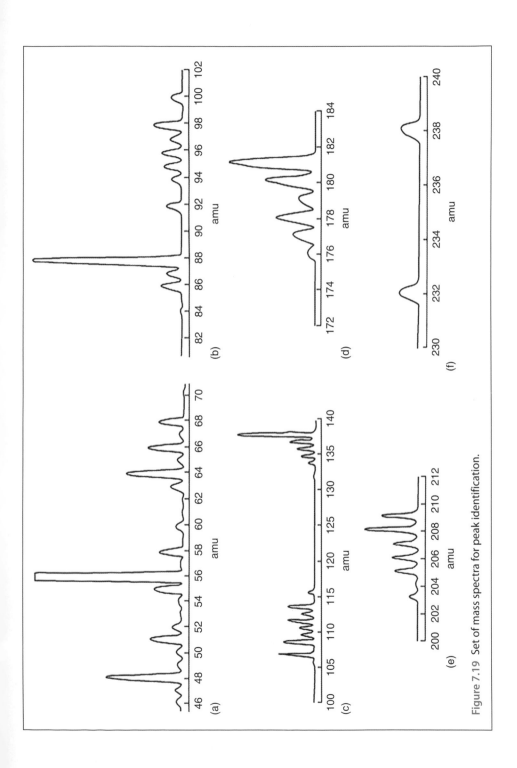

Figure 7.19 Set of mass spectra for peak identification.

Table 7.8 Relative abundances of selected naturally occurring isotopes [14], with data taken from Lide [15].

Symbol/amu	46	47	48	49	50	51	52	53	54	55	56	57	58	59	60	61	62	63	64	65	66	67	68	69	70
Ti	8.0	7.3	73.8	5.5	5.4																				
V					0.2	99.8																			
Cr					4.3		83.8	9.5	2.4																
Mn										100															
Fe									5.9		91.7	2.1	0.3												
Ni													68.1		26.2	1.1	3.6		0.9						
Co														100											
Cu																		69.2		30.8					
Zn																			48.6		27.9	4.1	18.8		0.6

Symbol/amu	82	83	84	85	86	87	88	89	90	91	92	93	94	95	96	97	98	99	100	101	102	103	104
(Kr)[a]	11.6	11.5	57.0		17.3																		
Rb				72.2		27.8																	
Sr			0.5		9.9	7.0	82.6																
Y								100															
Zr									51.4	11.2	17.1		17.4		2.8								
Nb												100											
Mo											14.8		9.3	15.9	16.7	9.5	24.1		9.7				

Symbol/amu	105	106	107	108	109	110	111	112	113	114	115	116	117	118	119	120	121	122	123	124	125	126
Ru														5.5		1.9	12.7	12.6	17.1	31.6		18.6
Rh																					100	
Pd																				1.0		11.1
Pd	22.3	27.3		26.5		11.7																
Ag			51.8		48.2																	
Cd		1.2		0.9		12.5	12.8	24.1	12.2	28.7		7.5										
In									4.3		95.7											
Sn								1.0		0.6	0.4	14.5	7.7	24.2	8.6	32.6		4.6		5.8		
Sb																	57.4		42.6			
Te																0.1		2.6	0.9	4.7	7.1	18.8
I																						
Xe																				0.1		0.1

Symbol/amu	128	129	130	131	132	133	134	135	136	137	138	139	140	141	142	172	173	174	175	176	177	178
Te	31.7		34.1																			
Xe	1.9	26.4	4.1	21.2	26.9		10.4		8.9													
Cs						100																

(Continued)

Table 7.8 (Continued)

Symbol/amu	128	129	130	131	132	133	134	135	136	137	138	139	140	141	142	172	173	174	175	176	177	178
Ba			0.1		0.1		2.4	6.6	7.9	11.2	71.7											
La											0.1	99.9										
Ce									0.2		0.3		88.4		11.1							
(Yb)																21.9	16.1	31.8		12.7		
Lu																			97.4	2.6		
Hf																		0.2		5.2	18.6	27.3

Symbol/amu	179	180	181	182	200	201	202	203	204	205	206	207	208	209	232	233	234	235	236	237	238
Hf	13.6	35.1																			
Ta		0.01	99.99																		
(W)		0.1		26.5																	
(Hg)					23.1	13.2	29.9		6.9												
Tl								29.5		70.5											
Pb									1.4		24.1	22.1	52.4								
Bi														100							
Th															100						
U																		0.7			99.3

a) Symbols in parentheses indicate elements that have incomplete percentage abundances (<100%).

Table 7.9 Results from mass spectral interpretation.

(a) The mass spectrum contains Ti, V, Cr, Mn, Fe, Ni, Cu and Zn.

Symbol/amu	46	47	48	49	50	51	52	53	54	55	56	57	58	59	60	61	62	63	64	65	66	67	68	69	70
Ti	8.0	7.3	73.8	5.5	5.4																				
V					0.2	99.8																			
Cr					4.3		83.8	9.5	2.4																
Mn										100															
Fe									5.9		91.7	2.1	0.3												
Ni													68.1		26.2	1.1	3.6		0.9						
Cu																		69.2		30.8					
Zn																			48.6		27.9	4.1	18.8		0.6

(b) The mass spectrum contains Sr and Mo.

Symbol/amu	82	83	84	85	86	87	88	89	90	91	92	93	94	95	96	97	98	99	100	101	102	103	104
Sr			0.5		9.9	7.0	82.6																
Mo											14.8		9.3	15.9	16.7	9.5	24.1		9.7				

(Continued)

Table 7.9 (Continued)

(c) The mass spectrum contains Ag, Cd and Ba.

Symbol/amu	105	106	107	108	109	110	111	112	113	114	115	116	130	131	132	133	134	135	136	137	138
Ag			51.8		48.2																
Cd		1.2		0.9		12.5	12.8	24.1	12.2	28.7		7.5									
Ba													0.1		0.1		2.4	6.6	7.9	11.2	71.7

(d) The mass spectrum contains Hf and Ta.

Symbol/amu	173	174	175	176	177	178	179	180	181	182
Hf		0.2		5.2	18.6	27.3	13.6	35.1		
Ta								0.01	99.99	

(e) The mass spectrum contains Tl, Pb and Bi.

Symbol/amu	200	201	202	203	204	205	206	207	208	209	210
Tl				29.5		70.5					
Pb					1.4		24.1	22.1	52.4		
Bi										100	

(f) The mass spectrum contains Th and U.

Symbol/amu	231	232	233	234	235	236	237	238	239
Th		100							
U					0.7			99.3	

7.8 Summary

The use of the inductively plasma for mass spectrometry is highlighted, with the fundamentals of mass spectrometry, as related to its combination with an ICP, being discussed. The major spectrometer designs for ICS–MS are described. Emphasis is placed on the occurrence of interferences in mass spectrometry and the potential remedies for overcoming them. An important alternative strategy that is offered using mass spectrometry, namely IDA, is also described.

References

1 Gray, A.L. (1974). *Proc. Soc. Anal. Chem.* 11: 182–183.
2 Houk, R.S., Fassel, V.A., Flesch, G.D. et al. (1980). *Anal. Chem.* 52: 2283–2289.
3 Gray, A.L. (1985). *Spectrochim. Acta* 40B: 1525–1537.
4 Gray, A.L. (1986). *J. Anal. At. Spectrom.* 1: 403–405.
5 Bogaerts, A. and Aghaei, M. (2017). *J. Anal. At. Spectrom.* 32: 233–261.
6 Morita, H., Ito, H., Uehiro, T., and Otsuka, K. (1989). *Anal. Sci.* 5: 609–610.
7 Bradshaw, N., Hall, E.F.H., and Sanderson, N.E. (1989). *J. Anal. At. Spectrom.* 4: 801–803.
8 Rehkamper, M., Schonbachler, M., and Stirling, C.H. (2000). *Geostand. Geoanal. Res.* 25: 23–40.
9 Vanhaecke, F., Balcaen, L., and Malinovsky, D. (2009). *J. Anal. Atom. Spectrom.* 24: 863–886.
10 Bolea-Fernandez, E., Balcaen, L., Resano, M., and Vanhaecke, F. (2017). *J. Anal. Atom. Spectrom.* 32: 1660–1679.
11 US Patent US2143262 A (12th March 1935), P.T. Farnsworth, Means for electron multiplication.
12 Tanner, S.D., Baranov, V.I., and Bandura, D.R. (2002). *Spectrochim. Acta* 57B: 1361–1452.
13 Koppenaal, D.W., Eden, G.C., and Barinaga, C.J. (2004). *J. Anal. At. Spectrom.* 19: 561–570.
14 Jakubowski, N., Moens, L., and Vabhaecke, F. (1998). *Spectrochim. Acta* 53B: 1739–1763.
15 Lide, D.R. (ed.) (1992/1993). *CRC Handbook of Chemistry and Physics*, 73e, 11-28–11-132. Boca Raton, FL: CRC Press.

8

Inductively Coupled Plasma: Current and Future Developments

LEARNING OBJECTIVES
• To compare the key figures of merit of ICP–AES and ICP–MS. • To be aware of the diverse field of application of the ICP with AES or MS. • To be aware of current and future developments, for the ICP, for ICP–MS and for ICP–AES. • To provide some guidance on other useful resources. • To provide templates for practical laboratory use.

8.1   Introduction

Often, the choice of analytical technique, inductively coupled plasma-atomic emission spectrometry (ICP–AES) versus inductively coupled plasma-mass spectrometry (ICP–MS), is not relevant. This is because most laboratories may not have both to perform elemental analyses. They might have other analytical techniques that can perform elemental analysis including electrochemistry (e.g. voltammetry), flame photometry (for Group I and II elements in the Periodic Table), X-ray fluorescence spectroscopy (for analysis of solid and liquid samples) or ion chromatography (for metal and anion analysis). Thus, overall the choice of analytical technique is far wider than presented in this book. Nevertheless, it is appropriate to consider the merits and disadvantages of both ICP–AES and ICP–MS techniques.

8.2   Comparison of ICP–AES and ICP–MS

One way to compare the ICP–AES and ICP–MS is via some important analytical figures of merit for each technique. However, this can only be done at a superficial level as the experience of the operator, the choice of manufacturer

Practical Inductively Coupled Plasma Spectrometry, Second Edition. John R. Dean.
© 2019 John Wiley & Sons Ltd. Published 2019 by John Wiley & Sons Ltd.

of the specific instrument and the frequency of its operation (by the user) all contribute to the actual reality of a comparison.

It might be considered that the most important parameters are:

- detection limits
- sample throughput
- analytical working range
- purchase and operating costs

A summary of the comparison of both techniques is made in Table 8.1. The definition of detection limit was defined earlier (Section 1.7, Eq. (1.3)) as the lowest amount of element in a sample that can be detected but not necessarily quantified as an exact value. A comprehensive list of detection limits for both techniques is shown in Table 8.2. It is observed that ICP–MS is far superior in terms of detection limits. This also brings with it other concerns to achieve these incredible low detection limits. Most notably, the requirement for the operation of the ICP–MS and the preparation of standard and samples for its application in use in a Clean Room. It is also important to be aware that impurities will be detected by the ICP–MS occur in everything that meets the prepared solutions, from the storage, measurement and transfer vessels used (and its previous cleaning), the dilution water (and its inherent impurities) and the reagent used (the analytical standard will never be 100% pure, nor the acid used in its preparation and solution stabilization). In other words, control of the physical working environment and the chemicals used are paramount. Practically, this means that reagent and sample blanks are mandatory, as well as strict attention to detail and the heightened possibility of cross-contamination.

The analytical working range is the concentration range over which the analytical working calibration plot remains linear; that is, the linear dynamic range (Table 8.1). This has the considerable advantage of allowing samples with different element concentrations to be determined without further dilution or pre-concentration, thereby saving analysis time. The analytical working range is expressed in terms of the order of magnitude of signal intensity, one order representing a factor of 10. In terms of these techniques, both have large analytical ranges with the possibility of analysing samples over six orders of magnitude. For the higher concentration range ICP–AES is more appropriate, while for the lowest concentration ranges ICP–MS is required.

Sample throughput is the number of samples that can be analysed or elements that can be determined in a specific time interval. As both techniques can be operated for simultaneous multi-element analysis, both have a high sample throughput. The techniques can process up to 40 elements per minute in a sample. The capital equipment purchase of either technique is high, with ICP–MS being the highest. Both techniques can utilize specific sample introduction features, for example, hydride generation, laser ablation and

Table 8.1 Comparison of the performance of ICP–AES and ICP–MS.

Criterion	ICP–AES	ICP–MS
Detection limits	Sub-ppb – ppm	Sub-ppt
Analytical capability	Multi-element	Multi-element
Sample throughput	Approx. 1–5 mins/sample	Approx. 1–4 mins/sample
Linear dynamic range	High ppm range	Mid ppb range
Precision		
Short term	0.1–2%	0.5–2%
Long term	1–5%	<4% (4 h)
Interferences		
Spectral	Some	Few
Chemical	Very few	Some
Physical	Some	Some
Elements applicable to	>60	>80
Sample volume required	1–2 ml min^{-1}	0.02–2 ml min^{-1}
Capital cost of instrument	High cost (~£70 K); depends whether simultaneous (higher cost) or sequential.	Higher cost (>£100 K).
Instrument operating costs (excluding normal supply of electricity and water, as required)	Requires large quantities of argon gas. Periodic replacement of ICP torch.	Requires large quantities of argon gas. Periodic replacement of ICP torch and sample/skimmer cones.
Ease of use of instrument	Operation is relatively straightforward based on a nebulizer/spray chamber. Obviously, complexity is added with other sample introduction devices. Software skills required for operation.	Operation is relatively straightforward based on a nebulizer/spray chamber. Obviously, complexity is added with other sample introduction devices. Software skills required for operation.
Possibilities for automation	Autosampler used.	Autosampler used.
Cost per sample (overall)	Medium	Medium

Source: Adapted from 'Guide to Inorganic Analysis', Perkin-Elmer, Inc., 2004.

chromatography, which add to the overall capital cost. However, both techniques are capable of multi-element analyses, with high sensitivity and the ability to measure more than one element in a sample. It should also be noted that the operating costs of inductively coupled plasma (ICP) techniques are not insignificant due, in part, to the high consumption of argon.

Table 8.2 Detection limits for elements using atomic spectroscopic techniques ($\mu g\ l^{-1}$).

Element symbol	Element	ICP-AES[a]	ICP–MS[b],[c]
Ag	Silver	0.6	0.00009
Al	Aluminium	1	0.0004*
As	Arsenic	1	0.0004
Au	Gold	1	0.0001
B	Boron	1	0.001
Ba	Barium	0.03	0.00004
Be	Beryllium	0.09	0.003
Bi	Bismuth	1	0.00002
Ca	Calcium	0.05	0.0003*
Cd	Cadmium	0.1	0.00007
Ce	Cerium	1.5	0.00005
Co	Cobalt	0.2	0.00006*
Cr	Chromium	0.2	0.0003*
Cs	Caesium		0.00005
Cu	Copper	0.4	0.0002*
Dy	Dysprosium	0.5	0.0002
Er	Erbium	0.5	0.0001
Eu	Europium	0.2	0.00007
Fe	Iron	0.1	0.0005*
Ga	Gallium	1.5	0.00008
Gd	Gadolinium	0.9	0.0003
Ge	Germanium	1	0.0006*
Hf	Hafnium	0.5	0.0003
Hg	Mercury	1	0.001
Ho	Holmium	0.4	0.00004
In	Indium	1	0.00008
Ir	Iridium	1	0.00009
K	Potassium	1	0.0001
La	Lanthanum	0.4	0.00004
Li	Lithium	0.3	0.00005
Lu	Lutetium	0.1	0.00004
Mg	Magnesium	0.04	0.0001

Table 8.2 (Continued)

Element symbol	Element	ICP-AES[a]	ICP–MS[b],[c]
Mn	Manganese	0.1	0.0001*
Mo	Molybdenum	0.5	0.0008
Na	Sodium	0.5	0.0003
Nb	Niobium	1	0.00004
Nd	Neodymium	2	0.0003
Ni	Nickel	0.5	0.0002*
Os	Osmium	6	0.00006
P	Phosphorus	4	0.04*
Pb	Lead	1	0.00004*
Pd	Palladium	2	0.00003
Pr	Praseodymium	2	0.00003
Pt	Platinum	1	0.0001
Rb	Rubidium	5	0.0002
Re	Rhenium	0.5	0.0003
Rh	Rhodium	5	0.00004
Ru	Ruthenium	1	0.0001
S	Sulfur	10	0.9*
Sb	Antimony	2	0.0002
Sc	Scandium	0.1	0.001
Se	Selenium	2	0.0003*
Si	Silicon	10	0.09
Sm	Samarium	2	0.0002
Sn	Tin	2	0.0002
Sr	Strontium	0.05	0.00007
Ta	Tantalum	1	0.00001
Tb	Terbium	2	0.00003
Te	Tellurium	2	0.0003*
Th	Thorium	2	0.00005
Ti	Titanium	0.4	0.0002*
Tl	Thallium	2	0.00001
Tm	Thulium	0.6	0.00003
U	Uranium	10	0.00002

(Continued)

Table 8.2 (Continued)

Element symbol	Element	ICP-AES[a]	ICP-MS[b),c]
V	Vanadium	0.5	0.00007*
W	Tungsten	1	0.00003
Y	Ytrrium	0.2	0.00002
Yb	Ytterbium	0.1	0.0001
Zn	Zinc	0.2	0.0007*
Zr	Zirconium	0.5	0.00007

a) Simultaneous multi-element conditions used with axial viewing of a dual-view plasma using a cyclonic spray chamber and a concentric nebulizer.
b) All ICP–MS measurements were made on a NexION ICP-MS with a quartz sample introduction system using a three second integration time and 10 replicates in de-ionized water. Detection limits were measured under multi-element conditions in Standard mode, except where denoted by an asterisk (*). Detections denoted by * were performed in a Class-100 Clean Room using Reaction mode with the most appropriate cell gas and conditions for that element in de-ionized water.
c) All detection limits are based on a 98% confidence level (three standard deviations).
Source: Adapted from 'Atomic Spectroscopy – A Guide to Selecting the Appropriate Technique and System', Perkin-Elmer, Inc., 2008–2013.

8.3 Applications

Both techniques (ICP–AES and ICP–MS) can be applied to a whole range of applications that require elemental information. The techniques are applied in a diverse and growing application field that allows their inherent benefits to be applied. Typically, the techniques might be applied across a broad range of applications and uses from the routine to cutting-edge research and this will encapsulate product development, quality control, drug discovery, food safety and so on in a diverse range of industries (e.g. in the metals, chemicals, advanced materials, petroleum and nuclear industries; environmental regulation and monitoring; clinical and biological materials; foods and beverages; archaeology and geochronology and geology).

Some not so obvious examples include:

- Environmental analysis.
- Analysis of halogenated (e.g. chlorine, bromine, or iodine) volatile organic compounds; organophosphorus and organochlorine pesticides (e.g. diazinon or DDT) using gas chromatography-inductively coupled plasma-mass spectrometry (GC–ICP–MS).
- Chemical warfare agents (e.g. organophosphorus nerve agent degradation products) using reversed phase-high performance liquid chromatography-inductively coupled plasma-mass spectrometry (RP–HPLC–ICP–MS).

- Flame retardants (e.g. polybrominated diphenyl ethers, PBDE congeners) using GC–ICP–MS.
- Petroleum analysis (e.g. speciation of sulfur [thiophenes] compounds) using GC–ICP–MS.
- Clinical and biological materials.
- Thyroid hormones (triiodothyronine (T_3) and thyroxine (T_4)) detection of iodine by RP–HPLC–ICP–MS.
- Analysis of phospholipids in biological samples by HPLC–ICP–MS.
- Identification and analysis of methyl-selenium metabolites by HPLC–ICP–MS.
- Phosphoric acid triesters in human plasma by solid phase microextraction – GC–ICP–MS.
- Phosphorylation profiling of tryptic protein digests using capillary LC–ICP–MS.
- Analysis of copper and zinc containing superoxide dismutase by isoelectric focusing gel electrophoresis laser ablation–ICP–MS.

Each application often has its unique sample preparation approach that allows the sample to be introduced in to the ICP in an appropriate form. In addition, the coupling of a 'front-end' accessory to the ICP can allow, for example, species-specific information to be assessed. An example of this could be the use of HPLC–ICP to allow the determination of organometallic species in food, plant and biological samples. Alternatively, the use of laser ablation-ICP will allow, for example, geological prospecting for precious metals to take place.

A unique feature of ICP–MS, compared to ICP–AES, is its ability to measure isotopes. This ability can be used specifically to quantify elements that have more than one isotope using an approach called isotope dilution analysis (IDA). IDA can then be applied, for example, in tracer-studies relating to plant, clinical and biological materials. For IDA to be used, an element must have more than one isotope. Then, by adding an artificially enriched isotope of the same element to the sample and measuring isotope ratios, the concentration of the element in the original sample can be determined.

In terms of robustness ICP–AES can withstand a wider range of matrices, samples with high salt contents and organic solvent-containing samples compared to ICP–MS due to the absence of a significant 'interface' consisting of a sample cone and skimmer cone (see Section 7.2.2).

8.4 Current and Future Developments

The ICP as a source for AES and mass spectrometry is relatively new and modern. The speed of innovation in new technologies, computing power and scientific discovery means that developments that were unheard of at the start of the

1990s as well as the 2000s are now commonplace. And so are the current and future developments in the ICP. This section tries to outline some of these current developments as well as speculate on future developments.

8.4.1 In General, for the ICP

- Automated, unattended and robotic sample handling and preparation for ICP. The ability to detect low element concentrations (see Table 8.2) requires further consideration. These additional considerations include the purity of the reagents (and water) used to prepare the samples, the sample container itself, the working environment for sample preparation and instrument operation (the requirement for a Clean Room). Therefore, without special measures and procedures the impact on the analytical data through elevated blank concentration and the probability of cross-contamination are high. The further development of robotic sample handling in a Clean Room would also reduce the possibility of cross-contamination and speed up productivity.
- Continuous development and evolution of sample introduction arrangements seek to improve transport of the sample into the ICP in a range of contexts. Examples include the use of temperature- (Peltier-) controlled spray chambers to reduce the solvent loading into the plasma, ultrasonic nebulization, chemical and photochemical (Figure 8.1) [1] generation to improve sample transport efficiency for As, Bi, Br, Cd, Co, Fe, Hg, I, Ni, Os, Pb, Se, Sb, Sn and Te), alternate nebulizers (Blurring®) that generate aerosols with both small droplets and a narrow particle size distribution (Figures 8.2 and 8.3) [2–4], and the resurgence in interest in electrothermal vaporization, ETV (Figure 8.3) [5].

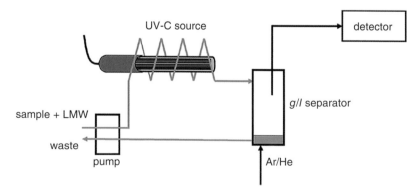

Figure 8.1 Photochemical vapour generation for ICP [1]. [*Notes: g/l* separator = gas liquid separator; LMW = low molecular weight (organic acids). Typical organic acids used include formic acid, acetic acid and propionic acid. UV–C = 280–100 nm spectral region. Enough photon energy for homolytic cleavage of, for example, C–C bonds.] Reproduced with permission of the RSC.

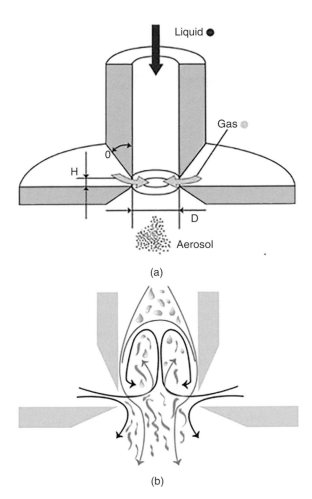

Figure 8.2 Schematic diagram of a (a) flow Blurring nebulizer and (b) its mode of operation. Reproduced with permission of Ingeniatrics. [*Notes*: In flow Blurring nebulization, turbulent mixing occurs between a liquid (blue) and gas (black). The flow Blurring nebulizer creates an aerosol with extremely fine micro- and nano-scale droplets as well as having high liquid compatibility and the potential to operate across a wide range of solution flow rates.] *Source*: Courtesy of Ingeniatrics.

8.4.2 For ICP–MS

- One of the most significant developments in ICP–MS instrumentation was the inclusion of collision-reaction cells (CRCs), to remove spectral interferences in quadrupole instruments. A further improvement in the removal of interferences was the introduction of the triple quadrupole inductively coupled plasma-mass spectrometry (ICP–QQQ–MS) (Figure 8.4). In the

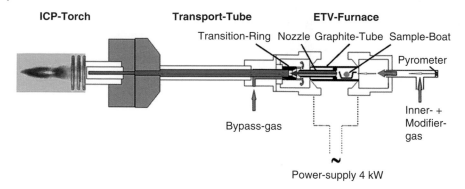

Figure 8.3 An electrothermal vaporization (ETV) sample introduction device for an ICP [5]. [*Notes:* Small sample aliquots are rapidly heated in the ETV furnace and the generated sample vapour is transported in a flow of gas directly to the ICP torch.] Reproduced with permission of the RSC.

ICP–QQQ–MS the first quadrupole eliminates ionized matrix interferences. As the analyte of interest enters the CRC, isobaric and multiple charged interferences can be removed (or the analyte reacted to a different m/z), while the final quadrupole filters the m/z ion of the analyte of interest.

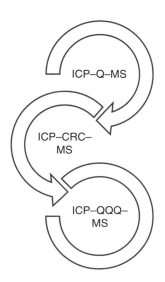

Figure 8.4 A significant development in ICP–MS to eliminate interferences.

A complex example would be the analysis and identification of Ti, using NH_3 in the CRC and monitoring the ion for $Ti(NH_3)_6^+$.

- Speciation studies and their implications for human health continue to develop. While traditionally high focus has been on arsenic/organoarsenicals, chromium(iii/vi), mercury/methylmercury and tin/organotins, so significant instrumental refinements have taken place. These include developments of the chromatographic coupling interface for gas chromatography (GC) and high performance liquid chromatography (HPLC) to the ICP, understanding and reduction of interferences (isobaric and polyatomic interferences) using inductively coupled plasma-collision-reaction cell-mass spectrometry (ICP–CRC–MS) and high resolution-inductively coupled plasma-mass spectrometry (HR–ICP–MS) (Figure 8.5). The field is therefore open for further investigations of other known toxic/non-toxic element combinations.

- The potential of high speed, simultaneous detection (e.g. ICP–ToF–MS) would lead to developments in bioimaging (i.e. mapping the distribution of metals in biological tissues or other natural or man-made materials). This is being enhanced by the development of improved laser ablation cells. In addition, the use of ICP–ToF–MS (Figure 8.6) with its fast scan speed is being applied in mass cytometry, an important tool in cell biology as well as cancer research. As the reporter molecules in mass cytometry are

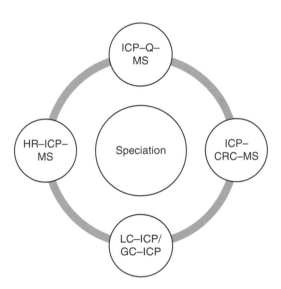

Figure 8.5 Developments in speciation studies because of innovation in ICP–MS technologies.

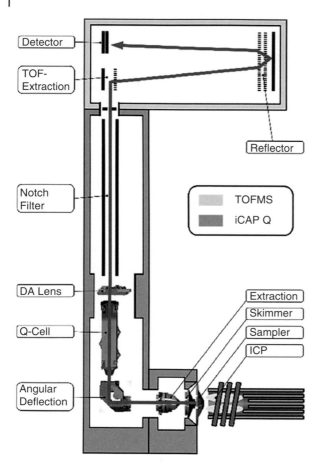

Figure 8.6 Schematic diagram of a commercial ICP–ToF–MS [6]. [*Notes*: The flight path of the ions is highlighted in red. The ions are extracted from the ICP interface and deflected through 90° that removes neutral species (from the ion beam). The ions are then focused into the collision cell (Q-cell) to which different gases can be added. The subsequent ions pass through the notch filter that selectively attenuates ions of up to four different *m/z* ratios. The remainder of the ion beam is focused into the extraction region of the orthogonal-acceleration time-of-flight (TOF) mass analyser.] Reproduced with permission of the RSC.

usually antibodies, it is possible to tag them with isotopically pure rare earth elements (REEs). The REE tagged antibodies then bind with specific cell components, and the mass cytometer then measures the expression of the antibodies in each individual cell. The major advantage of the use of REEs is their ability to be detected at much lower concentrations by

ICP–MS, rather than the traditional fluorophore-tagged antibodies used in flow cytometry.

- The release of engineered nanoparticles (e.g. Ag, Au, Pt, Pd, TiO_2) into the environment because of human health and environmental effects is of concern, as well as their monitoring and characterization during manufacturing. At this point, the high speed, simultaneous detection by ICP–ToF–MS is being explored.

- Geochronology, the science of determining the age of rocks, fossils and sediments, uses the element signatures inherent in the rocks themselves. Developments in multiple collector–inductively coupled plasma–mass spectrometry (MC–ICP–MS) have allowed precise isotopic element ratios to be determined, more rapidly and with simpler solution nebulization, laser ablation, hyphenated chromatographic separation and vapour generation, compared to thermal ionization mass spectrometry (TIMS), which operates under high vacuum (including the ionization stage). Also, while TIMS is restricted to elements with ionization energies (Table 7.1) ≥ 7.5 eV, this is not the case with an ICP. This advantage allows the determination, by ICP, of non-metals including the determination of the isotope ratios for bromine and sulfur.

- The diverse and expanding field of 'omics' is being explored by ICP–MS technologies. Specifically, metallomics, that focuses on the role of metals in biological systems: for metals that are depleted or in short-supply, that is, Cu, Fe, Mo or Zn, or in excess with harmful effects, that is, As and Cr.

8.4.3 For ICP–AES

- In the future, the replacement of charge coupled device (CCD) and charge-injection device (CID) detectors for ICP–AES with detectors based on complementary metal-oxide semiconductor (CMOS) circuitry. Currently, the technology is used to produce integrated circuits where they are used for microprocessors, batteries and digital camera image sensors. As they are currently being used as a replacement for the CCDs in camera sensors, it is only a matter of time before they will be applied in atomic emission spectrometry (AES).

- To overcome the self-absorption and matrix effects associated with axial viewing of the ICP for AES a synchronous vertical dual-view configuration, SVDV (Figure 8.7) has been developed. It provides combined simultaneous axial- and radial-viewing configurations. The SVDV configuration uses a dichroic spectral combiner that allows selection and combining of the

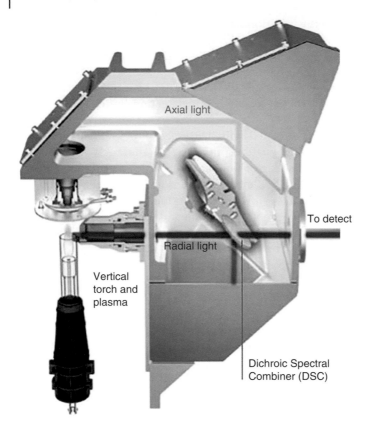

Figure 8.7 Synchronous vertical dual-view of a commercial ICP–AES [7]. [*Notes*: Emitted radiation from the axial and radial plasma views are synchronously converged onto the dichroic spectral combiner, prior to being transmitted to the spectrometer.] Reproduced with permission of the RSC.

radiation from both viewing positions simultaneously. This new viewing platform allows the simultaneous determination of major and minor elements.

Finally, we await the first commercial plasma-based system with a dual mass spectrometry (MS) detection system for both elemental and molecular species. A summary of the main developmental stages in the growth of ICP technologies is illustrated in Figure 8.8.

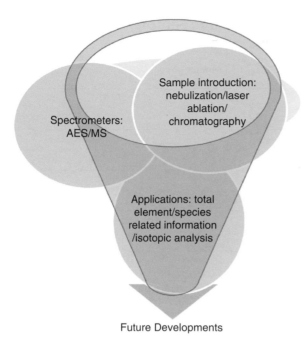

Future Developments

Figure 8.8 The main developmental stages in the growth of ICP technologies.

8.5 Useful Resources

A useful resource when using either instrument is the use of Laboratory Templates (Templates 8.1–8.4). Template 8.1 'Laboratory Template: Sample Pre-Treatment' allows the user to enter details of all the appropriate sample pre-treatments that could have been used to prepare a sample (solid, liquid or gaseous) for analysis. While Template 8.2, 'Laboratory Template: Sample Preparation' covers the specific details of how a sample will have been prepared prior to its analysis. Finally, Templates 8.3 and 8.4, cover the specific instrument settings for ICP–AES (Template 8.3) and ICP–MS (Template 8.4). All the templates act as a useful reminder of the level of detail required to be recorded when performing instrumental analytical chemistry.

Template 8.1 Laboratory Template: Sample Pre-Treatment

Your name:	Module code:

Title of Experiment:

Sample description and source:

Sample dried: Yes/No[#]	Oven or air-dried[#]	Temperature: °C	Duration: hr
Grinding: Yes/No[#]	Grinder used (model/type):		Duration: mins
Sieving: Yes/No#	Mesh size:		Duration: mins

Sample mixing: Yes/No[#]	Manual: Yes/No[#]	Mechanical: Yes/No[#]	Action: orbital, rocking or other …….

Sample storage:	Location:		Date In:
solid / liquid / gas[#]	Container type:	Fridge: °C	Date Out:

Pre-treatment: Yes/No[#]	pH adjustment Yes/No[#]	Addition of alkali: Yes/No[#]	Addition of acid: Yes/No[#]
	Name of acid/alkali/buffer[#]:		Volume added: mL

Other practical observations:

[#] delete as appropriate

Template 8.2 Laboratory Template: Sample Preparation

Your name:		Module code:		
Title of Experiment:				
Sample description and source:				
Sample weights	Sample	1	2	3
(accurately weighed i.e. record to four decimal places)	Weight of sample + vessel:	g	g	g
	Weight of vessel:	g	g	g
	Weight of sample:	g	g	g
Sample volume	Collected: mL	Filtered: Yes/No#	Mesh size:	
Acid digestion: Yes/No#	Type of vessel used:	Hot plate or other Yes/No: #. If other please specify:		
	Temperature controlled or not: Yes/No# °C	Type of acid used:	Concentration of acid:	
	Weight of sample used: g	Volume of acid used: mL	Heating duration: mins	
Other method of sample decomposition: please specify with details				
Quality Control: Yes/No#	Certified Reference Material (CRM): Yes/No#	Type of CRM:		
	Weight used: g	Volume used: mL		

Stock Solution(s):	Calibration Standards: µg/mL		
	(low to high; calculated using C1V1 = C2V2).		
Internal standard added: Yes/No#	Element / wavelength: / nm	Concentration: µg/mL	
Blanks prepared: Yes/No#	Matrix-matched: Yes/No#		
Other practical observations:			

delete as appropriate

Template 8.3 Laboratory Template: ICP–AES Analysis

Your name:	Module Code:		
Title of experiment:			
Operating conditions			
ICP make / model:	Spectrometer type:		
Forward Power kW	Generator frequency: Hz		
Viewing mode:	Axial or Radial* (delete as appropriate)	Observation height (radial view only): mm above load coil	
Argon gas flow rates	Outer (plasma): L/min	Intermediate (auxiliary): L/min	Injector (nebulizer): L/min
Sample introduction	Nebulizer type:	Spray chamber:	Other:
Sample Analysis and Calibration			

Element (symbol):	1.	2.	3.	4.
Wavelength (nm):				

Quantitation	Internal standard: Yes/No#	Element:	Wavelength (nm):	Concentration: µg/mL

Calibration Standards: μg/mL

(low to high; calculated using $C_1V_1 = C_2V_2$).

Number of calibration	Element	Linear dynamic	Correlation	Element	Linear dynamic	Correlation
standards:		range	coefficient		range	coefficient
Dilution / concentration factor:	1			3		
	2			4		

Other practical observations:

* Axial view is also known as end-on viewing; Radial view is also known as side-one viewing.

Template 8.4 Laboratory Template: ICP–MS Analysis

Your name:	Module Code:

Title of experiment:

Operating conditions

ICP make / model:	Mass Spectrometer type:

Forward Power kW	Generator frequency: Hz

Argon gas flow rates	Outer (plasma): L/min	Intermediate (auxiliary): L/min	Injector (nebulizer): L/min
Sample introduction	Nebulizer type:	Spray chamber:	Other:

Sample Analysis and Calibration

Element (symbol):	1.	2.	3.	4.
Mass/charge ratio (amu):				

Quantitation	Internal standard: Yes/No#	Element:	Mass/charge (amu):	Concentration: ng/mL

Calibration Standards: ng/mL
(low to high; calculated using C1V1 = C2V2).

Number of calibration standards:	Element	Linear dynamic range	Correlation coefficient	Element	Linear dynamic range	Correlation coefficient
Dilution / concentration factor:	1			3		
	2			4		

Other practical observations:

References

1 Sturgeon, R.E. (2017). *J. Anal. At. Spectrom.* 32: 2319–2340.
2 Almagro, B., Ganan-Calvo, A.M., and Canals, A. (2004). *J. Anal. At. Spectrom.* 19: 1340–1346.
3 Aguirre, M.A., Kovachev, N., Almagro, B. et al. (2010). *J. Anal. At. Spectrom.* 25: 1724–1732.
4 Pereira, C.D., Aguirre, M.A., Nobrega, J.A. et al. (2014). *Microchem. J.* 112: 82–86.
5 Hassler, J., Barth, P., Richter, S., and Matschat, R. (2011). *J. Anal. At. Spectrom.* 26: 2404–2418.
6 Hendriks, L., Gundlach-Graham, A., Hattendorf, B., and Gunther, D. (2017). *J. Anal. At. Spectrom.* 32: 548–561.
7 Donati, G.L., Amais, R.S., and Williams, C.B. (2017). *J. Anal. At. Spectrom.* 32: 1283–1296.

Further Reading

Selected other books on inductively coupled plasma spectroscopy.

Thomas, R. (2013). *Practical Guide to ICP-MS: A Tutorial for Beginners, Third Edition (Practical Spectroscopy)*, 3e. CRC Press.

Moore, G.L. (2012). *Introduction to Inductively Coupled Plasma Atomic Emission Spectrometry*, vol. 3. Elsevier Science.

Zachariadis, G. (2012). *Inductively Coupled Plasma Atomic Emission Spectrometry*. Gazelle Publishing.Selected other books on sample preparation.

Baranowska, I. (2015). *Handbook of Trace Analysis: Fundamentals and Applications*. Springer.

Dean, J.R. (2014). *Environmental Trace Analysis. Techniques and Applications*. Wiley.

Flores, E.M.M. (2014). *Microwave-Assisted Sample Preparation for Trace Element Determination*. Elsevier.

Pawliszyn, J. (ed.) (2012). *Comprehensive Sampling and Sample Preparation*. Elsevier Inc.

Van-Gajic, J. (2012). *Samples & Sample Preparation in Analytical Chemistry*. Nova Science Publishers Inc.

9

Inductively Coupled Plasma: Troubleshooting and Maintenance

LEARNING OBJECTIVES

- To consider the potential causes of poor instrument sensitivity, instrument precision and instrument accuracy.
- To consider tips to reduce autosampler issues and contamination.
- To consider tips to improve sample preparation.
- To consider how to unblock a blocked pneumatic concentric nebulizer, clean the spray chamber and the plasma torch.
- To consider what to do about plasma ignition problems.
- To shut down the instrument (at the end of the day).
- To be aware of the regular maintenance procedures required on a periodic basis.

9.1 Introduction

This final chapter will provide some guidance for troubleshooting as well as routine maintenance for your inductively coupled plasma (ICP) system. However, in order that maintenance is required some issues will have arisen as pointers to the problem. Typical issues include the following.

9.2 Diagnostic Issues

Poor instrument sensitivity. *Causes* include worn or 'flat' spots on peristaltic pump tubing, blocked nebulizer, blocked injector in torch, incomplete or poor system optimization, possible interferences. *Remedies* include replace worn peristaltic pump tubing, clean the nebulizer/injector, re-optimize the instrument and check for interferences via sample matrix, acids used or instrument generated.

Practical Inductively Coupled Plasma Spectrometry, Second Edition. John R. Dean.
© 2019 John Wiley & Sons Ltd. Published 2019 by John Wiley & Sons Ltd.

Poor instrument precision. Causes include worn or 'flat' spots on peristaltic pump tubing, beading in spray chamber, intermittent blocked nebulizer, air-leak in transfer tubing, misaligned torch, incomplete or poor system optimization, and memory effects from incomplete sample wash-out. *Remedies* include replace worn peristaltic pump tubing, clean the nebulizer/spray chamber, replace the tubing, re-optimize the instrument and for memory effects, check the rinse time used is adequate (30–40 seconds is typical) and use an acidified (matrix-matched) rinse solution.

Poor accuracy. Causes include worn or 'flat' spots on peristaltic pump tubing, blockages in nebulizer and/or torch, interferences, incomplete or poor system optimization, choice of internal standard, insufficient stabilization time prior to signal recording, incomplete sample digestion, no matrix matching and memory effects from incomplete sample wash-out. *Remedies* include checking sample digestion procedure and efficiency, is the digest stable and free of precipitates or suspensions? And check for potential contamination from other reagents or digestion equipment (include a reagent blank with every sample batch run).

9.3 Tips to Reduce…

9.3.1 Potential Autosampler Issues

- Length and condition of transfer tube from autosampler to ICP: you may need to shorten the tubing or increase the sample uptake delay time. Also, check for kinks in the tubing and replace as necessary.
- Ensure the autosampler probe is appropriate for the sample matrix: use a wider bore for high total suspended solids or viscous samples.
- Sample stability while waiting in uncovered racks: potential for dust ingress, sample evaporation and sample settling.

9.3.2 Contamination

The risk of contamination is prevalent in sample preparation and analysis.

- Check reagent purity, including its grade of reagent and certificate of analysis.
- Always re-seal the reagent container immediately after use.
- Other common contamination sources include: the reagent water, typically with a conductivity of $18.2\,M\Omega\,cm$; the cleaning solution used for the glassware; storage of the glassware pre-use; the risk from airborne dust in the laboratory; the use of powdered gloves (especially for Zn) for sample handling and contamination risk from (coloured) pipette tips (especially Cd, Cu, Fe and Zn).

9.4   Tips to Improve...

9.4.1   Sample Preparation

- Check that purchased standards are still within their 'use by' date.
- Use calibrated pipettes and class 'A' volumetric flasks for dilutions; periodically check accuracy and reproducibility of pipettes.
- Use de-ionized water (conductivity $18.2\,M\Omega\,cm$).
- Use serial dilutions for preparing low concentrations from 10 000 ppm stock solutions; that is, multiple stages.
- Prepare low concentration standards ppb ($ng\,ml^{-1}$) daily from a high concentration stock.
- Prepare low concentration standards ppm ($\mu g\,ml^{-1}$) weekly from a high concentration stock.
- Store standards in plastic vessels; this ensures better stability.
- Stabilize standards with acid; a low pH ensures better stability.

9.5   How To ...

9.5.1   Unblock a Blocked Pneumatic Concentric Nebulizer

- Remove the nebulizer from its mounting and visually examine under 20× or 30× magnification.
- If particles are wedged inside the nebulizer, carefully and gently tap the liquid input of the nebulizer. If the particles come loose, then repeat the tapping to allow the particle to progress to the exit orifice. Apply compressed gas (up to 30 psi or $20.7 \times 10^4\,Pa$) to assist and if required. Alternatively, backflush the nozzle using isopropyl alcohol.
- Solid material is present, but flow through the nebulizer is still possible. In this case, inject an appropriate solvent into the nozzle to dissolve the solid deposit and then remove the solvent with compressed gas.
- Solid material is present and the nebulizer is blocked. Gently heat the nebulizer at the point of the blockage and then carefully apply gas pressure at the sample input tube.
- The nozzle is encrusted with crystalline deposits. Immerse the nozzle in an appropriate rinse solution. Then, if necessary, apply compressed gas (up to 30 psi [$20.7 \times 10^4\,Pa$]) to the nozzle.
- No foreign matter is visible. Immerse the nozzle in hot, concentrated nitric acid and repeat as necessary.

9.5.2   Clean the Spray Chamber

If 'beading' or droplet formation forms on the internal walls of the spray chamber, it requires cleaning.

1) Sonicate in a detergent solution (with care) overnight.
2) Rinse it, allow it to dry and then refit.

9.5.3 Clean the Plasma Torch

When the ICP torch needs cleaning, follow these principles:

1) Do not sonicate.
2) Soak the torch in concentrated aqua regia (3:1, v/v $HCl:HNO_3$) overnight.*
3) Rinse and allow to dry before re-installing.

9.6 What to do About...

9.6.1 Plasma Ignition Problems

In most, the plasma will ignite first time. Occasionally, however, it does not; this is typically due to the presence of air. So, the following order of response is suggested:

1) Repeat the ignition step.
2) Has the argon cylinder been replaced recently? Check its grade. Try another cylinder.
3) Check all connections in the sample introduction system for cracks, loose fittings, missing or damaged items.
4) Check that the plasma and auxiliary gas connections are the right way around.
5) Check the alignment of the torch with respect to the radiofrequency (RF) load coil/plate.
6) Has the sample changed? For example, organic solvents require higher RF power.
7) Check the various on/off buttons.

9.7 Shut Down Procedure (At the End of the Day)

1) Aspirate acid rinse solution for a few minutes before shutting off the plasma. This helps to reduce sample deposition build-up inside the nebulizer/spray chamber after the final run of the day.

---

* For more persistent deposits, use a pipecleaner dipped in aqua regia to clean the injector tube. For salt deposits, rinse further with water. If this is unsuccessful, soak overnight in a detergent solution.

2) Extinguish the plasma. For inductively coupled plasma-atomic emission spectrometry (ICP–AES), switch off the chiller.
3) Release the pressure on the peristaltic pump tubing and ensure that the tubing is no longer stretched over the pump rollers.
4) Empty the waste vessel.
5) Leave the mains power on (unless local operating procedures dictate otherwise).

9.8 Regular Maintenance Schedule

- Daily
  - Inspect the torch for injector blockage/other damage.
  - Check the nebulizer for blockage/pulsation.
  - Inspect the peristaltic pump tubing for stretching or flatness.

- Weekly
  - Clean the torch (or earlier, if required).
  - Check the sample introduction tubing.
  - Inspect the ceramic cone (axial) or snout (radial) in ICP–AES; clean if required by sonicating in dilute detergent.
  - Inspect the sample cone in inductively coupled plasma–mass spectrometry (ICP–MS); clean if required by sonicating in dilute detergent.
  - Inspect torch bonnet for cracks or sample deposition.
  - Wipe down the exterior surfaces of the instrument (especially the sample compartment).

- Monthly
  - Clean the spray chamber (earlier, if beading is evident).
  - Clean the nebulizer.
  - Check the sample introduction tubing: be vigilant for excessive wear, poor sealing or kinks and replace as necessary; in particular, check the transfer tube from spray chamber to the torch and spray chamber waste outlet. Replace or empty as necessary.
  - Inspect/clean the ceramic cone (ICP–AES in an axial configuration).
  - Inspect/clean the bonnet and/or snout on the torch (ICP–AES in radial configuration).
  - Inspect/clean the sampling and skimmer cone (ICP–MS).
  - Inspect the state of the induction coil/plate.
  - Clean/check the filter on the water chiller/recirculatory.
  - Clean/check the water level in the water chiller/recirculatory and top up as required.
  - Check the oil level in the vacuum pump (ICP–MS).

- Periodically (6–12 months)
    - Replace the water in the water chiller and dose with algaecide.
    - Replace the oil in the vacuum pump (ICP–MS).
    - Change argon filters on the argon gas supply (if using gas cylinders).

The Periodic Table

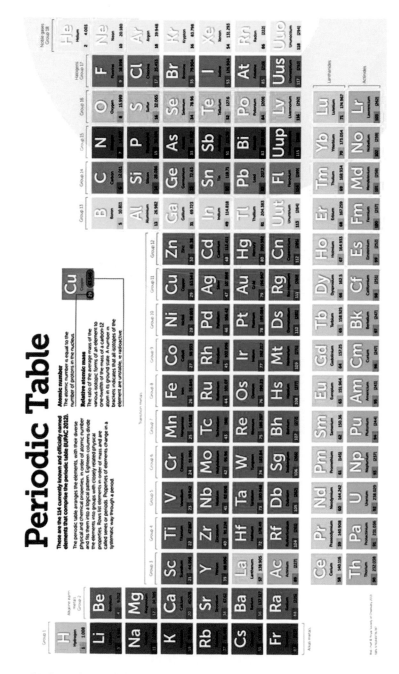

Practical Inductively Coupled Plasma Spectrometry, Second Edition. John R. Dean.
© 2019 John Wiley & Sons Ltd. Published 2019 by John Wiley & Sons Ltd.

SI Units and Physical Constants

SI Units

The SI system of units is generally used throughout this book. It should be noted, however, that according to present practice, there are some exceptions to this, for example, wavenumber (cm^{-1}) and ionization energy (eV).

Base SI units and physical quantities

Quantity	Symbol	SI unit	Symbol
length	l	metre	m
mass	m	kilogram	kg
time	t	second	s
electric current	I	ampere	A
thermodynamic temperature	T	kelvin	K
amount of substance	n	mole	mol
luminous intensity	I_v	candela	cd

Prefixes used for SI units

Factor	Prefix	Symbol
10^{21}	zetta	Z
10^{18}	exa	E
10^{15}	peta	P
10^{12}	tera	T
10^{9}	giga	G

Practical Inductively Coupled Plasma Spectrometry, Second Edition. John R. Dean.
© 2019 John Wiley & Sons Ltd. Published 2019 by John Wiley & Sons Ltd.

Factor	Prefix	Symbol
10^6	mega	M
10^3	kilo	k
10^2	hecto	h
10	deca	da
10^{-1}	deci	d
10^{-2}	centi	c
10^{-3}	milli	m
10^{-6}	micro	μ
10^{-9}	nano	n
10^{-12}	pico	p
10^{-15}	femto	f
10^{-18}	atto	a
10^{-21}	zepto	z

Derived SI units with special names and symbols

Physical quantity	SI unit		Expression in terms of base or derived SI units
	Name	Symbol	
frequency	hertz	Hz	$1\,\text{Hz} = 1\,\text{s}^{-1}$
force	newton	N	$1\,\text{N} = 1\,\text{kg}\,\text{m}\,\text{s}^{-2}$
pressure; stress	pascal	Pa	$1\,\text{Pa} = 1\,\text{Nm}^{-2}$
energy; work; quantity of heat	joule	J	$1\,\text{J} = 1\,\text{Nm}$
power	watt	W	$1\,\text{W} = 1\,\text{J}\,\text{s}^{-1}$
electric charge; quantity of electricity	coulomb	C	$1\,\text{C} = 1\,\text{As}$
electric potential; potential difference; electromotive force; tension	volt	V	$1\,\text{V} = 1\,\text{J}\text{C}^{-1}$
electric capacitance	farad	F	$1\,\text{F} = 1\,\text{CV}^{-1}$
electric resistance	ohm	Ω	$1\,\Omega = 1\,\text{VA}^{-1}$
electric conductance	siemens	S	$1\,\text{S} = 1\,\Omega^{-1}$
magnetic flux; flux of magnetic induction	weber	Wb	$1\,\text{Wb} = 1\,\text{Vs}$
magnetic flux density	tesla	T	$1\,\text{T} = 1\,\text{Wb}\,\text{m}^{-2}$

Physical quantity	SI unit		Expression in terms of base or derived SI units
	Name	**Symbol**	
magnetic induction inductance	henry	H	$1\,H = 1\,Wb\,A^{-1}$
Celsius temperature	degrees Celsius	°C	$1°C = 1\,K$
luminous flux	lumen	lm	$1\,lm = 1\,cd\,sr$
illuminance	lux	lx	$1\,lx = 1\,lm\,m^{-2}$

Derived SI units with special names and symbols

Physical quantity	SI unit		Expression in terms of base or derived SI units
	Name	**Symbol**	
activity (of a radionuclide)	becquerel	Bq	$1\,Bq = 1\,s^{-1}$
absorbed dose; specific energy	gray	Gy	$1\,Gy = 1\,J\,kg^{-1}$
dose equivalent	sievert	Sv	$1\,Sv = 1\,J\,kg^{-1}$
plane angle	radian	rad	$1^{a)}$
solid angle	steradian	sr	$1^{a)}$

a) rad and sr may be included or omitted in expressions for the derived units.

Physical Constants

Recommended values of selected physical constants[a]

Constant	Symbol	Value
acceleration of free fall (acceleration due to gravity)	g_n	$9.806\,65\,m\,s^{-2\,b)}$
atomic mass constant (unified atomic mass unit)	m_u	$1.660\,540\,2(10) \times 10^{-27}\,kg$
Avogadro constant	L, N_A	$6.022\,136\,7(36) \times 10^{23}\,mol^{-1}$
Boltzmann constant	k_B	$1.380\,658(12) \times 10^{-23}\,J\,K^{-1}$
electron specific charge (charge-to-mass ratio)	$-e/m_e$	$-1.758\,819 \times 10^{11}\,C\,kg^{-1}$

Constant	Symbol	Value
electron charge (elementary charge)	e	$1.602\,177\,33(49) \times 10^{-19}$ C
Faraday constant	F	$9.648\,530\,9(29) \times 10^4$ C mol^{-1}
ice-point temperature	T_{ice}	273.15 K[b]
molar gas constant	R	$8.314\,510(70)$ J K^{-1} mol^{-1}
molar volume of ideal gas (at 273.15 K and 101 325 Pa)	V_m	$22.414\,10(19) \times 10^{-3}$ m^3 mol^{-1}
Planck constant	h	$6.626\,075\,5(40) \times 10^{-34}$ J s
standard atmosphere	atm	$101\,325$ Pa[b]
speed of light in vacuum	c	$2.997\,924\,58 \times 10^8$ ms^{-1}[b]

a) Data are presented in their full precision, although often no more than the first four or five significant digits are used; figures in parentheses represent the standard deviation uncertainty in the least significant digits.
b) Exactly defined values.

Index

Practical Inductively Coupled Plasma Spectrometry, Second Edition. John R. Dean.
© 2019 John Wiley & Sons Ltd. Published 2019 by John Wiley & Sons Ltd.